Tropical Visions in an Age of Empire

Edited by Felix Driver and Luciana Martins

Tropical Visions in an Age of Empire

The University of Chicago Press   Chicago and London

Felix Driver is professor of human geography at Royal Holloway, University of London. *Luciana Martins* is lecturer in Luso-Brazilian Studies at Birkbeck, University of London.

The University of Chicago Press, Chicago 60637
The University of Chicago Press, Ltd., London
© 2005 by The University of Chicago
All rights reserved. Published 2005
Printed in the United States of America

14 13 12 11 10 09 08 07 06 05   1 2 3 4 5

ISBN: 0−226−16471−3 (cloth)
ISBN: 0−226−16472−1 (paper)

Library of Congress Cataloging-in-Publication Data

Tropical visions in an age of empire / edited by Felix Driver and Luciana Martins.—1st ed.
    p. cm.
Includes bibliographical references and index.
ISBN 0-226-16471-3 (cloth : alk. paper)—ISBN 0-226-16472-1 (pbk. : alk. paper)
    1. Tropics—Description and travel.   2. Europeans—Travel—Tropics.
3. Traveler's writings, European.   4. Scientific expeditions—Tropics.
5. Tropics—In art.   I. Driver, Felix.   II. Martins, Luciana.
G905. T75 2005
910′.913 — dc22

                                        2005001424

⊗ The paper used in this publication meets the minimum requirements of the American National Standard for Information Sciences—Permanence of Paper for Printed Library Materials, ANSI Z39.48−1992.

Contents

SITES

AFTERWORD

Illustrations

Acknowledgments

This book has its origins in a research project concerned with the ways in which European travelers apprehended tropical nature in the eighteenth and nineteenth centuries. Focusing principally on the visual archive of tropical travel, this project explored the relationship between the visualization of tropical landscape and the process of circulation through the tropics. In the course of the research, we became increasingly interested in the complex historical geographies of tropicality and the extent to which the tropical imaginary was shaped by particular regional formations within the Caribbean, Latin America, Africa, South Asia, East Asia, and the Pacific. *Tropical Visions in an Age of Empire* significantly expands the geographical range and historical scope of the original project, embracing a variety of sites across, and beyond, the tropical world. The book brings together authors form a variety of disciplines, including literature, cultural geography, art history, and the history of science in order to address the visualization of the "tropical" within a range of aesthetic, scientific, and political projects. We gratefully acknowledge the support of the Arts and Humanities Research Board for the original project. We would also like to take this opportunity to record our thanks to some of the many individuals with whom we have discussed various aspects of the ideas explored in these pages, in particular, David Arnold, Tim Barringer, Dipti Bhagat, Gavin Bowd, Michael Bravo, Dan Clayton, Denis Cosgrove, Phil Crang, Stephen Daniels, Paulo Geyer, Michael Heffernan, Peter Hulme, Tariq Jazeel, Bernhard Klein, David Lambert, Nigel Leask, David Lehmann, David Livingstone, David Lowenthal, Gesa Mackenthun,

Nancy Naro, Miles Ogborn, Rebecca Preston, Geoff Quilley, Hugh Raffles, Kapil Raj, Nigel Rigby, James Sidaway, Charles Withers, and Brenda Yeoh.

Initial versions of most of the essays were presented at a two-day conference at the National Maritime Museum, Greenwich, convened jointly with Nigel Rigby and Margarette Lincoln. We are very grateful to the staff of the museum for supporting the conference so magnificently and, in particular, to Janet Norton for her work on our behalf. We would also like to thank other contributors to the conference, including Harriet Guest, Dian Kriz, Nancy Stepan, Nicholas Thomas, and Beth Tobin, and the large number of participants whose contributions have been invaluable in the process of preparing this book for publication. The success of the conference also depended on the support of the British Academy, the Historical Geography Research Group, the National Maritime Museum, and the Royal Geographical Society, which we gratefully acknowledge here.

In producing and revising this book for publication, we have enjoyed the benefit of detailed comments on the entire manuscript from the readers for the University of Chicago Press. We thank the contributors for their patience and good humor in the process of seeing the book through to publication. We are particularly grateful to Denis Cosgrove for agreeing to write the afterword and doing so in such style, and to Starr Douglas for compiling the index. Finally, we would like to record our heartfelt thanks to Christie Henry, Jennifer Howard, and Yvonne Zipter for their editorial advice and assistance at every stage of the project.

INTRODUCTION

Views and Visions of the Tropical World

FELIX DRIVER AND LUCIANA MARTINS

Our knowledge of the earth is constructed in a variety of ways, through experience, learning, memory, and imagination. This book is devoted to an exploration of images of the tropical world produced by European travelers over the past three centuries. It is concerned more particularly with the ways in which tropical places are encountered and experienced, the significance of travel for the process of producing knowledge about these places, and the relationship between geographical difference and generalized notions of "tropicality." The contrast between the temperate and the tropical is one of the most enduring themes in the history of global imaginings. Whether represented positively (as in fantasies of the tropical sublime) or negatively (as a pathological space of degeneration), tropicality has frequently served as a foil to temperate nature, to all that is modest, civilized, cultivated. The idea of the tropical as a distinct assemblage of natural and human relations has taken diverse forms in different geographical and intellectual settings. *Tropical Visions in an Age of Empire* brings together contributors from various disciplinary backgrounds— principally art history, cultural geography, literature, imperial history, and the history of science—in order to consider the visualization of the tropical world.

Images of tropical natures and cultures have a long history and complex geography.[1] At different moments, in different contexts, the notion of tropicality has been enrolled in a variety of philosophical, political, scientific, and aesthetic projects. Within the literatures of natural history, travel, and exploration, for example, the idea of tropical difference has had a remarkably sustained influence, even when—perhaps especially when—

the actual experience of tropical travel has failed to live up to expectations. Throughout the early modern period, its presence can also be detected in a host of cultural forms, from epic poetry to landscape painting, as well as in the historical and philosophical reflections on human nature and the wealth of nations. From the nineteenth century, we see tropical difference given institutional expression in the emergence and development of distinct subdisciplinary specializations—tropical medicine, tropical climatology, tropical geography, and so on—though in each of these fields the definition and limits of the "tropical" have been anything but settled. Over the past century, the discourse on tropicality has further proliferated under the influence of modernism, decolonization, development discourse, global tourism, commodity advertising, and environmental politics. The tropics, then, have long been the site for European fantasies of self-realization, projects of cultural imperialism, or the politics of human or environmental salvage. In the postcolonial world, these fantasies have if anything become more pervasive, if distinctly less enchanting.

The imaginative flow has certainly not at all been one way. Artists and intellectuals working in what we now call the global South have appropriated the language of tropicality for their own ends, and this in turn has influenced the ways in which Europeans have understood tropical nature and culture. Take, for example, the concept of Luso-tropicalism, initially developed by the Brazilian sociologist Gilberto Freyre during the 1930s. In his writings, Freyre emphasized the ideal of the harmonious blending of racial, religious, and cultural differences that he suggested had emerged historically in Brazil during the colonial period. His conception was subsequently adopted by the governing elite within Portugal in an attempt to provide an ideological framework that would sustain what remained of their colonial empire in Africa and Asia.[2] Within Brazil itself, Freyre's concerns with cultural fusion would later be reframed by the aesthetic of tropical modernism, as, for example, in the work of the landscape architect Roberto Burle Marx.[3] Further encounters between modernism and tropicality were reflected in the ambivalent cultural politics of the *Tropicália* movement in Brazilian popular music during the late 1960s, most notably in the work of Caetano Veloso.[4] These are of course highly specific engagements with the idea of tropicality, in place, style, and period. But in these cases the task of interpretation leads us beyond rigid distinctions between the metropolis and colony, core and periphery, the cosmopolitan and the indigenous, temperate and tropical, as we see new forms of tropicality emerging in the process of transculturation.[5]

The conventional discourse of tropicality might be compared with that of Orientalism, to the extent that both have conventionally been used to define and legitimize essential differences between cultures and natures, both understood in strongly spatial terms.[6] However, our emphasis on cultural encounter and exchange prompts questions about the ways in which the effects of such discourses are often conceived. In particular, the model of projection that drives some accounts of colonial discourse, the "West" projecting its sense of cultural difference on the "rest," needs to be problematized.[7] One obvious risk here is that images (like "the Orient" or "tropicality") are conceived as already fully formed, ready-to-be-projected, a position that greatly exaggerates their coherence and consistency. Another is that the cultural and natural worlds of the East, or the tropics, are represented as homogenous screens on which these images of difference are depicted. A properly postcolonial perspective must also bring into question not just the representation of Europe's others but also the production of "Europe" itself—in our case, Europe as a space of temperate culture and nature. We have become so used to thinking of European expansion—including the exploration and colonization of the tropical world—as the means of extending and dramatizing an already existing worldview, that we have underestimated the extent to which the process of extension is actually transformative of the European sense of culture and history—of the temperate self. Culturally as well as economically speaking, this European self has never been self-sufficient: it has always learned, borrowed, or stolen from elsewhere.[8] We need to develop ways of conceiving this process of exchange in terms of transactions rather than projections: to think of images, certainly, but to understand the process of their being made as negotiated—and sometimes contested—in various ways.[9] This would enable the production of knowledge about the tropical world to be understood as a more differentiated, more uneven, and ultimately more human process; and moreover, it would give more agency, and autonomy, to the world being represented—understood not simply as a screen but as a living space of encounter and exchange.[10]

TROPICAL VIEWS AND VISIONS

In this volume, we are especially concerned with the ways in which "the tropics" have been represented as something to be seen—a view to be had or a vision to be experienced. In the writings of European naturalists during the eighteenth and nineteenth centuries, notably Alexander von Humboldt and Alfred Russel Wallace, tropical nature was figured in strikingly visual terms. During the same period, traveling artists, such as William Hodges and Johann Rugendas, attempted to give form to the new sense of tropicality, which was emerging in the course of European scientific exploration. The significance of tropical views and visions for the development of new models of science is explicit in the case of Humboldt,

FIGURE 1.1 Friedrich Georg Weitsch, *Alexander von Humboldt,* 1806, oil on canvas (Staatliche Museen zu Berlin, Alte Nationalgalerie)

who once declared the tropics to be "his element."[11] His lyrical depictions of tropical nature and its sublime geographies inspired many subsequent travelers, not least Charles Darwin. Arriving in Rio de Janeiro in April 1832, Darwin wrote of his first sight of "a tropical forest in all its sublime grandeur," as if the scene demanded such a response from any truly philosophical traveler. "I formerly admired Humboldt, I now almost adore him; he alone gives any notion of the feelings that are raised in the mind of first entering the Tropics."[12] Humboldt's writings and reputation loomed large over the discourse of tropicality during the nineteenth century and indeed right up to the present day (fig. 1.1). The reasons for renewed scholarly attention to his work in recent years are not hard to find. It raises far-reaching questions about the relationship between science and aesthetics, about the balance between holistic and analytical views of nature, and about the prospect of reconciling sedentary scholarship with observation in the field. This book seeks to situate Humboldt's influential vision in the larger context of European encounters with the tropical world.

As the essays in this volume attest, the spaces of the tropics have been imagined in a wide variety of ways, within diverse forms of writing, sketching, mapping, charting, panoramas, painting, and photography. In very general terms, we use the words "view" and "vision" here to capture two contrasting modalities through which the tropics have been pictured. The view emerged in the context of a topographical aesthetic, in which landscapes are depicted at a distance, their surface features translated into a recognizable visual code. In this very general sense, the term belongs equally to landscape sketching, coastal survey, and terrestrial mapping: it is part of a topographic culture in which the world is apprehended from afar. The vision, in contrast, is something that in principle takes hold of the observer in a much more transformative way: it engages the imagination and turns the spectator into an active participant in the scene. Where the view is the product of an enlightened reason, the vision is the means of asserting a new sensibility: the realization not just of an image of the world but of a new sense of self as well. In this sense, Humboldt's vision of the tropical world is rather unlike, say, Cook's views, insofar as it brings the eye of the observer itself into the frame. Of course, this distinction is more about epistemology than practice, as the fantasy of eliminating all traces of subjectivity from the map or the chart could never be fully realized. The lines on the paper spoke of desire as well as distance; in the case of maritime surveyors, for example, they trace an experience of trial, error, and inference—and in particular, speculation about what lay beyond the visible coastline.[13]

In this context, Humboldt's significance lies in his efforts to synthesize views and visions within a new conception of the natural world. In recent years, much attention has been devoted to his interests in various forms of visual representation as a means of apprehending the complex unity of nature, his perpetual return to reflections on the aesthetics of landscape, and the uneasy relationship between the personal narrative and the scientific overview in his writings.[14] Humboldt's speculations on the sources of human enjoyment in the contemplation of the face (*Ansichten,* or "aspects") of nature suggested that such an aesthetic had a history—and a geography. In a different register, as Michael Dettelbach shows in this volume, his own brand of self-experimentation as a traveling observer was designed to measure the effects of landscape impressions on his own sensibility. Humboldt reflected further on these themes in *Cosmos,* where he suggested that the distinctive physiognomy of the tropical landscape was expressed in a specifically tropical aesthetic, associated above all with sentiments of grandeur and luxuriance. "In the Humboldtian version of romantic imagination," Nigel Leask suggests, "aesthetic and emotional responses to natural phenomena counted as data about these phenomena, in contrast to their rigorous exclusion from contemporary practices of naval and military surveying."[15]

In accounting for the spirit of Humboldtian science, or the efforts of artists to capture the physiognomy of tropical nature, it is tempting to exaggerate the constancy and consistency of the European gaze in the tropics. This book pays particular attention to the ways in which travel has the capacity to dislodge the certainty of the self, confronting European travelers in the tropics with often unexpected visions. If knowledge of the tropical world was not always a settled knowledge but frequently a contradictory knowledge-in-the-making, how should historians view the documents that are its material traces? Here we have an opportunity to reconsider approaches to the visual inventory of tropical travel and the variety of interpretative strategies it allows. According to the model of cultural critique founded on the idea of projection, Europeans often saw what they wanted to see when they traveled into the tropics, projecting an imaginative geography of natural and cultural difference onto the new worlds they encountered. But tropical nature and society was far more than a screen, and the apparently simple act of projection was in fact a laborious process in which a variety of transactions were involved. Thinking in these terms enables us to conceive the work of representation as a process of unequal exchange, suggesting an alternative point of departure for historians of tropicality.[16] It also allows for a more discriminating view of the coherence of the European view of the tropics, one in which the experience of disorientation, uncertainty, and novelty has its place.[17] This in turn raises questions about the multiple practices through which the tropics were known, practical and bodily as well as intellectual and discursive; and it encourages greater attention to the ways in which European conceptions of the tropics may have been shaped by interactions with a wide variety of peoples and places.

VOYAGES

The modern cartographic definition of the tropics is rooted in the astronomical, climatic, and moral geographies of antiquity, in which the habitable earth or *oecumene* is identified as that portion of the globe lying between the torrid and frigid realms. The torrid or tropical zone is bounded by two parallels of latitude stretching around the earth, one 23°27′ north of the equator and the other 23°27′ south, together marking the limits of the region in which the sun shines directly overhead. In this cosmographical vision, the circles of Capricorn and Cancer define both a natural and a moral limit: the intertropical zone is imagined as a realm of otherness, beyond humanity. The genealogy of the "monstrous races" at the ends of the earth as represented in *mappae mundi* and medieval encyclopedias, as Denis Cosgrove has shown, owes much to this classical vision.[18] The expansion of the Ptolemaic *oecumene* in the long sixteenth century produced new visions of otherness and a newly historicized vision of European destiny: in the frontispiece to Ortelius's *Theatrum orbis terrarum* of 1570, for example, a crowned Europe is seated above figures representing Asia, Africa, America, and the much dreamed-of southern continent.[19] The wonders of the "new world" that dominate this age of discovery were accommodated within reworkings of both classical geographical theory and biblical schemas: significantly, this is a moment of continuity as well as transformation. In this context, too one sees signs of more familiar tropes of tropicality, as in Sir Walter Raleigh's account of his quest for El Dorado, in which the "indecipherable landscape" of Guiana is figured alternately as a plentiful paradise and an unrelenting hell.[20]

The role of eighteenth- and nineteenth-century voyages of exploration in perpetuating or challenging dominant views and visions of the tropical world is a major theme in this context. The new planetary consciousness of science was reflected in the global scale and functions of maritime expeditions of navigators like Cook and La Pérouse, the ambition of terrestrial explorers like Humboldt and Park, and the efforts of metropolitan savants and statesmen who sought to make the world an orderly place in the name of enlightenment.[21] They imagined the establishment of vast archives of texts, images, artifacts, and specimens, patiently assembled, through which the geography and natural history of the earth could be made known. They created great empires of learning presided over in Britain, for example, by such influential figures as Joseph Banks, Roderick Murchison, and Joseph Hooker, whose networks extended across every continent and every sea. Theirs was a suitably imperial vision, of order, system, and progress, in which the explorer's role was to fill in the blanks: the keepers of the imperial archive would do the rest. Looking more closely at the archive of tropical travel, however, it is clear that such projects raised as many questions as they answered. Could the encounter with the tropics challenge as well as confirm European constructions of nature? How was the experience of traveling itself to be put into words and images? To what extent did the experience of encountering difference, in nature and culture, undermine existing canons and conventions? Such questions were first seriously addressed in Bernard Smith's seminal work on the impact of Pacific island cultures and landscapes on the development of European scientific theories and landscape art between the ages of Cook and Darwin.[22] Smith's interpretations of the work of William Hodges, who accompanied Cook on his second Pacific voyage, are extended here in Claudio Greppi's chapter on traveling artists.[23] If *European Vision and the South Pacific* remains a fundamental reference point today, even for quite different forms of analysis and interpretation, it is partly because of its concerns with the epistemological status of image making—in what ways, precisely, can seeing be the equivalent of knowing?—and partly because of its treatment of the experience of travel and encounter, in which the space of experience is left open.[24]

In order to address the mediated nature of imaginative geographies, in this case those associated with tropicality, it is important to take seriously the spaces of experience and encounter. The archive of tropical exploration and navigation yields evidence of something more fragile and unpredictable than representations of planetary consciousness: in a word, disturbance.[25] Jonathan Lamb, referring to eighteenth-century voyages of discovery, puts the case well: "The commanders of these expeditions may have been committed to large and comprehensive views, and believed devoutly in systems of classification and cadastral measurement; but their data proved intractable, their experiments prone to failure, and they became periodically distracted, behaving unlike themselves owing to the stress of isolation, disease, fear—and occasionally exquisite pleasure."[26] The fantasy of constructing an inventory of global nature came at a cost: creating knowledge "on the spot" and transmitting it in a manner that was acceptable at home was far from a straightforward task. The claims of returning travelers were liable to be greeted with incredulity, especially if they lacked the credentials of the gentleman-philosopher. At the same time, navigators and explorers wrote contemptuously of "armchair geographers" speculating in the comfort of their salons and clubs. Such disputes are not merely colorful episodes in an otherwise straightforward history of enlightenment: they point to wider questions about the epistemology and authority of science in the field, which became particularly acute when that field lay within the torrid zone. As Dorinda Outram has emphasized, the literature on tropical travel raised troubling questions about the status of the explorer's knowledge.[27] From the perspective of the sedentary natural philosopher, scientific mastery depended less on the bodily experience of movement into new spaces than on the observer's very capacity to stand at a distance. Also, unlike the cabinet or the laboratory, the field was necessarily a more public space, inhabited by a wide range of people with diverse interests. Observation in the field accentuated a problem that the practice of science necessarily engendered: the question of trust. This was a problem that the deployment of new and more precise instruments could only partially resolve. It also required that travelers conduct their observations and their writings, and indeed present themselves, in a manner deemed appropriate to the protocols of scientific inquiry in the field. A whole methodology of observance, then, was designed to ensure that reliable and unvarnished information could be collected, stored, and eventually transmitted back to the center.

In this context, it is important to emphasize that questions of epistemology—how the tropics became known, both locally and globally—are more than purely cognitive in nature. As well as abstract concepts and techniques, the production and transmission of knowledge about tropicality involved a visceral engagement with particular places and sites: this applies also to the production of visual images in situ, given the skills, techniques and materials required in the making of views "on the spot." The empirical knowledge of the tropics that emerges from the later decades of the eighteenth century relied on a new conception of the body as a recording instrument, experiencing through the senses what it meant to be within the tropical, as opposed to speculating on it at a distance. It thus makes little sense to divorce questions of representation—the depiction of tropical nature—from questions of experience—dwelling within tropical nature. Here it is not so much a matter of pitting the "actual" realities of tropical travel against the "perceived" images of tropical landscapes, or the practical as against the representational engagement with landscape, as of considering the entanglement between these different registers through which the tropical world became known. The problem of representing tropical difference turns into another, more fundamentally geographical question. How does being in the tropics affect one's knowledge of it?

MAPPINGS

If the mapping of the world was central to the projects of Enlightenment thought, materially and metaphorically, it was because mapping involves, in the most general sense, translation through inscription.[28] To map entails taking the measure of the world, observing its phenomena, and then figuring its data so that it may be accessible to others—whether in cartographic, diagrammatic, textual, or some other form.[29] In the context of Enlightenment thought, mapping is one way of reconciling an interest in difference—the study of human and natural variations across the globe—with a commitment to uniformity in both method and ultimately explanation. In this sense, the mapping of the topography, hydrography, geography, natural history and ethnology of the tropical world posed a series of problems for European travelers. How were established procedures and principles of mapping to comprehend cultures and landscapes that were fundamentally unfamiliar? How were their characteristic forms to be translated into a language that could be recognized? To what extent did the sheer otherness of tropicality—its climate, vegetation, and landforms, as much as its human geographies—place it in a world apart?

Historians of science are unanimous in their estimation of the role of mapping within the program of Humboldtian science, even where they disagree about the relationship between its aesthetic and analytical components. Many have argued that Humboldt's concerns with the geographical distribution of natural phenomena and its graphic representation, notably in the form of iso-maps, provide the key to his scientific reasoning and its influence on others, including Darwin: in essence, according to

FIGURE 1.2 Alexander von Humboldt, *Géographie des plantes près de l'Equadeur,* 1803, ink and watercolor on paper (Museo Nacional de Colombia, Bogotá)

Susan Cannon, Humboldtian science presented "a *topographical vision* of the world, its organisms, and its history."[30] In this context, it might be noted that works like Maury's *Physical Geography of the Sea,* discussed here by Graham Burnett, were conceived in much the same spirit: this was one face of a Humboldtian *physique du globe,* dedicated to the measurement, mapping and analysis of relationships between terrestrial, oceanic, and atmospheric phenomena. One of Humboldt's most celebrated works based on his tropical travels, the *Essai sur la géographie des plantes* (published in Paris in 1805), reflected his concern with both the variability and the unity of landscape, notably in the elaborate cross-sectional profile of the Andes that accompanied it. In this *tableau physique,* tropical botany is mapped both spatially and analytically: the landscape diagram, combining picture and text denoting the names of plants found at various altitudes, with a table of sixteen columns alongside, allows relationships between all manner of meteorological topographical, geological, botanical, and physical variables to be seen in one all-embracing view. Figure 1.2 shows the original sketch that provided the basis for Humboldt's well-known diagram, said to have been composed at the foot of Chimborazo, then supposed to be the highest mountain in the world. Humboldt here effectively synthesizes a global conception of tropicality, defined in the horizontal plane as it were, with a vision of the mountain sublime. The tropical location is vitally important: at this spot, the true variability and order of the entire natural world become visible, condensed on the vertical scale of elevation.[31]

Humboldt's distinction between the descriptive methods of the *botaniste nomanclateur* and the analytical concerns of the *botaniste physicien,*

elaborated in his essay on the geography of plants, rested essentially on the
ability of the latter to identify and in principle to map spatial variations in
the distribution of natural phenomena and the relationships between such
distributions. In the words of Michael Dettelbach,

> What was important was that physics was conceived of as essentially map-
> ping work, as generating lines by moving through real, physical space with
> precise instruments. Such lines demonstrated that there was, after all, in the
> necessary and unbridgeable gap between the artifice of mathematics or
> geometry and bodily, sensual, physical nature, between the calculus (a per-
> fect but empty language) and real human language (opaque with the den-
> sity of history, custom, and sensation), a shape of nature, which could at
> least in principle be more or less adequately drawn.[32]

Humboldt's celebrated *tableau* is an ingenious development of something
rather more commonplace in the literature of travel and exploration during the eighteenth and nineteenth centuries: that is, the combination of
graphic representations with other kinds of data, including descriptions of
manners and customs. As Michael Bravo has argued, the topographic maps
of such influential geographers as James Rennell provided a powerful tool
for a systematic and comparative science: "Reading signs on the *surface* of
the landscape (the sources of rivers, oases, cloud patterns) provided the key
for piecing together the landscape's inner propensities for imperial commerce — the direction and flow of its waters, the moral qualities of its populations and the caravan routes for the traffic in humans."[33] As the example of Rennell illustrates, the culture of precision in the contemporary
discourse of travel and exploration was not confined to the use of refined
instruments or numerical data: it was also reflected in new approaches to
evidence in the form of maps, travel narratives, and oral testimony.[34] Recent work on the survey of India during this period has emphasized not
only the role of the comparative methods of map compilation in the
offices of the East India Company but also the continuing reliance on local negotiations with Indian personnel and their surveying skills.[35]

The theme of mapping was also prominent in comparative studies of
ethnology from the second half of the eighteenth century on. At the
global scale, variations in the physical and cultural forms of humanity were
measured and accounted for in a variety of ways, notably through stadial
theories of human development, in which geographical differences could
be mapped onto the plane of linear time. Here the scale of civilization depended broadly on the level of cultivation and trade, at least as far as Europeans could recognize it, from hunting societies, through pastoral, agri-
cultural, commercial and industrial. However, the data arising from the
literature of scientific travel created the possibility of cultural mapping

at the regional and local scales. Johann Reinhold Forster, naturalist on Cook's second voyage, developed a more fine-grained comparative ethnology in his accounts of the "varieties of the human species" within the South Pacific, published in *Observations Made during a Voyage Round the World* in 1778. In his account, as Nicholas Thomas points out, a variety of aesthetic, moral, and political criteria are used to differentiate between the "nations" of Oceania, including physical appearance, civility, and the status of women. Significantly, Thomas detects a tension between two strands of Forster's argument, as he tries to reconcile his accounts of local variations with more general theories of human development. "The first strand of Forster's argument draws the societies observed into a particularizing, historical geography; the second assimilates them to a general narrative of political evolution."[36] One of the challenges facing historians of tropical encounters is that of teasing out the hesitancies and uncertainties that may have generated such tensions—not only in published narratives but also in sketches, portraits, and paintings.[37]

The suggestion that cultural stasis might be the natural counterpart to tropical bounty has ancient roots. And from the age of Columbus onward, the contrast between the productivity of tropical nature and the supposed absence of enterprise among its original inhabitants was routinely used to serve a colonial purpose.[38] The subsequent development of plantation economies, worked by forced labor, was of course justified partly on these grounds. The associated transfer of peoples and plants within the tropical zone helped to establish the tropics as a space of circulation, connecting Africa and the Americas, India and Southeast Asia, the Pacific and the Caribbean. Initially pioneered by the Portuguese, the Dutch, the French, and the British later followed suit with more developed schemes. As this suggests, the complex genealogy of notions of tropicality reveals a complex pattern of relation and mutation, within and beyond Europe: there is little space here for national exceptionalism. In the British case, a network of botanical gardens was established in support of schemes for plant exchange designed to rationalize the economic resources of empire, connecting Kew with Jamaica and Saint Vincent in the West Indies, Saint Helena in the South Atlantic, Calcutta and Madras in India.[39] In this context, it should be recalled that the purpose of the ill-fated voyage of the aptly named *Bounty* in 1788–89, commanded by William Bligh with the patronage of Sir Joseph Banks, was to transfer a cargo of breadfruit seedlings from the South Pacific to the West Indies. There, the breadfruit was to be transplanted in order to provide food for plantation slaves, a cheaper alternative to imported grain. Bligh eventually made the journey successfully, along with the breadfruit, in HMS *Providence,* in 1792–93. George Tobin, a young lieutenant who sailed with Bligh on this voyage, depicted

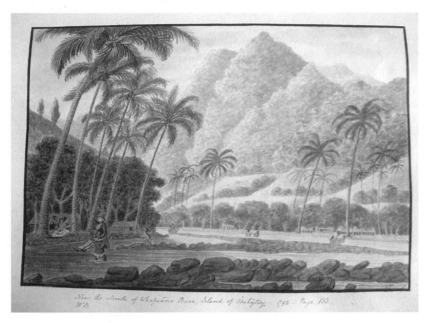

FIGURE 1.3 George Tobin, *Near the Mouth of Whapiano River, Island of Otahytey,* 1792, watercolor sketch (State Library of New South Wales, Sydney)

FIGURE 1.4 George Tobin, *Blue Fields, Jamaica,* 1793, watercolor sketch (State Library of New South Wales, Sydney)

the landscapes of Tahiti and Jamaica in a series of watercolor sketches that owed much to the tradition of coastal survey while providing a picturesque record of tropical ports of call (figs. 1.3 and 1.4). The geography of Tobin's sketch books reiterates the itineraries through which plants, people, and ideas were circulating across the tropical world, providing a visual correlate to the comparative trajectory of theories of racial difference examined in Peter Hulme's chapter in this volume.[40]

SITES

The discourse on tropicality is characterized more often by typification than generalization: in the process, very particular views and visions, represented, for example, by iconic images of tropical forest or desert island scenes, stand in for tropical landscapes as a whole. The iconography of tropical nature has been the subject of considerable attention in recent years, most notably in the context of sublime visions of tropicality in the work of naturalists and artists.[41] The process by which images of race, climate, plant, or landscape were translated from particular sites, located in particular regions, and produced as generic of the tropical world as a whole tells us much about the production and circulation of imaginative geographies. However, regional or local variations within the tropical world might also be highlighted, suggesting that this process of typification could operate at different geographical scales, depending on the context. Alfred Russel Wallace, for example, used conspicuous biogeographical variations within the Malay Archipelago to make his case for natural selection—hence the famous "Wallace line." In a different setting, the supposed dichotomy between the peoples of "Polynesia" and "Melanesia" has served a variety of purposes within geographical and ethnological writings since the eighteenth century.[42]

The conceptual apparatus of tropicality itself has a geography. As David Arnold has pointed out, the idea of the tropics as a distinctive environmental and cultural region was heavily influenced by the experience of travel to and settlement in the Americas and the Caribbean, notably in the context of natural history and medicine. It was through the literature of medical geography, well developed in this context, that the language of tropicality was eventually extended to the Indian subcontinent in the early nineteenth century. This perceptual shift, Arnold argues, was expressed through two rather different imaginative geographies. On the one hand, the India of the plains could be aligned with images of the hot, dry climate of the Middle East and its political correlate, Oriental despotism. On the other, there was the India of Bengal, whose humid climate was associated by the British with moral and physical weakness, contrasting with more "manly" races of the northwest. In an aside, Arnold highlights what he calls "an important piece of intra-tropical semantic exchange": while the term "hurricane" had traveled from the Caribbean to the East Indies, India was responsible for the word "jungle," which notwithstanding its original Sanskrit and Hindi meanings came to signify dense damp forests throughout the tropics.[43] Much more could be said about such exchanges, which reflect patterns of circulation through and beyond the tropical world. In the context of views and visions, for example, it is clear that representations of specific places within the tropical and subtropical world often bore the imprint of a variety of different sites. If the forests of the Americas could thus be imaginatively transported to the foothills of the Himalayas, so too could Oriental scenery be mapped onto the topography of Rio de Janeiro.[44] Even within Europe itself, the glare of the Mediterranean sun could provide the pretext for exploiting the "tropical" connotations of Southern landscapes.[45] The rhetoric and iconography of tropicality was far more hybrid than has hitherto been acknowledged.

European colonial expansion in Africa, the Pacific, and Southeast Asia in the late nineteenth and early twentieth centuries had a marked impact on the ways in which the tropics were represented. This is also the period in which knowledge of the tropical world received institutional expression in the formation of new disciplines, such as tropical medicine, which were closely allied with the requirements of colonial policy, including newly revived questions of acclimatization.[46] It is important to note here that while the distinction between the spaces of the tropical and the temperate was in some respects reinforced, it remained an uncertain boundary. As Rod Edmond suggests in his chapter, the formation of fields such as tropical medicine did not reflect a consensus on, for example, whether particular diseases were confined to tropical regions or even whether thinking of the "tropical world" as a whole was useful to science: the same might be said, later, of tropical geography. The appeal to tropicality continued to provide a powerful imaginative foundation for a variety of scientific, aesthetic, and political projects, and one, moreover, on which careers could be built. Alongside the professor of scientific fields such as tropical medicine, another figure emerges—the pioneer of tropical modernism, entranced by the fantasy of escape from the corruption of a settled metropolitan life in the quest for a new art. "What I want to do is to set up a studio in the *Tropics*" announced Paul Gauguin in 1890, prior to his departure for Tahiti. "I will perhaps become the Saint John the Baptist of the painting of the future, invigorated there by a more natural, more primitive, and, above all, less spoiled life."[47] Gauguin, like Humboldt, felt himself in his element. More significantly, in this context, he too made the tropics what it has now become. Today, tropical tourists of a certain class may cruise the South Pacific aboard the *Paul Gauguin,* a luxury vessel "expressly designed to harmonize with the beauty of the South Seas and the richness of their culture." [48]

★   ★   ★

The remaining essays in this book are organized into three sections: voyages, mappings, and sites. These represent three of the key modalities through which tropical views and visions are made: first, the oceanic and terrestrial voyages, through which tropical natures and cultures come to be known and represented; second, the processes and practices of mapping, through which particular places and zones across the globe are compared, contrasted, and made intelligible within a general framework; and third, the representation and negotiation of particular sites, in both the tropical and the temperate zones, associated with particular aspects of tropicality. The first essay, by Claudio Greppi, provides an overview of the work of traveling artists between the mid-eighteenth and mid-nineteenth centuries, beginning with William Hodges, who accompanied Cook on his second Pacific voyage and subsequently worked in India. The efforts of Hodges and others to depict tropical landscape had a significant impact on European visions of the tropics, especially as mediated through the work of Alexander von Humboldt. Humboldt's keen reflections on landscape painting and the aesthetics of landscape observation were in turn appropriated by a new generation of traveling artists. The result was a way of seeing, and knowing, in which the tradition of landscape art was fused with a new spirit of observation informed by the experience of voyaging around the world in the company of naval surveyors, meteorologists, and astronomers. In Humboldt's case, this emergent epistemology of landscape was accompanied by a new sensibility, which is the focus for Michael Dettelbach's essay. Humboldt's preoccupations with tropical travel are connected here to a late Enlightenment culture of sensibility. Humboldt—and those who followed in his footsteps, many quite literally—saw the tropics as a site for enacting a new model of the self, which visions of tropical nature helped to sustain. The notion of philosophical travel required something more than mere collecting becomes clear in the case of a naturalist such as William Burchell, the subject of our essay in this section. Here the focus is on Burchell's attempt to capture the pattern of tropical nature—not only in his collections of natural objects such as plants and animals but also in his creation of proxy specimens in the form of precise calibrated drawings. For Burchell, as for Humboldt, the art of visual representation—the depiction of nature's forms—was a vital tool of scientific description.

The second section considers some of the ways in which the diversity of the tropical world has been mapped, across islands, continents, and oceans. Peter Hulme considers how the space of the tropics was constructed as a space of comparison and circulation, using the examples of two very different islands: Dominica and Tahiti. Both islands have often been described as exemplary sites of tropicality, through a series of ethnographic, geographical, and moral tropes that ran closely in parallel. The mapping of tropicality did not simply work in parallel, however. By the late eighteenth century, as can be seen, for example, in the voyages of Captain Bligh, the Pacific and the Caribbean had been brought into the same space, as far as the circulation of people and plants were concerned. The intellectual framework for thinking comparatively about the human geography of tropical races soon followed. The themes of comparison and circulation are also considered in the chapter by Starr Douglas and Felix Driver, which examines the visual mapping of tropical nature through the work of the naturalist Henry Smeathman. In the course of his travels in Sierra Leone and the Caribbean during the 1770s, Smeathman effectively followed the route of the triangular trade, using this experience to draw parallels and contrasts between tropical nature in both its "rude" and its "cultivated" state. Smeathman's spectacular sketches of termite colonies provided a different kind of mapping of tropical nature, resulting in a composite image of landscape simultaneously picturesque, topographic, and analytical in form. In Graham Burnett's chapter, the focus turns from land to sea, to the hydrographic mapping of Matthew Fontaine Maury. Examining Maury's uncertain place in the history of science, Burnett considers the significance of his mapping methods for the development of oceanography and for his particular conception of tropicality. Maury's *Physical Geography of the Sea,* it is argued, lay firmly in the tradition of natural theology. But far from being an obstacle, this provided a congenial frame for his Humboldtian vision of science.

In the third section, the essays engage with three geographical sites, highlighting the ways in which they become articulated with particular concerns about aspects of tropicality. David Arnold considers Joseph Hooker's journey to India in 1848 and the extent to which it was initially inspired and subsequently described in terms of a conventional repertoire of tropical views. Hooker's prior apprenticeship as a naturalist and, especially, his experience of life at sea were instrumental in shaping his vision of the Indian landscape. His expectations of tropical scenery had a variety of literary and scientific sources, though his actual experience of travel through India often resulted in expressions of disappointment. The Himalayan foothills, with their combination of tropical and temperate flora, brought forth a variety of more Humboldtian associations. For Hooker, tropicality was something to be written about as much as measured or mapped, and his *Himalayan Journals,* as Arnold shows, reflected the aesthetic as well as the scientific impulses within Victorian natural history. Whereas Arnold's focus is principally on narrative and, to a lesser extent, graphic representation, Leonard Bell considers photography as a source of powerful images of tropical exoticism often associated (in Europe and North America) with the South Seas in general and Samoa in particular. In his chapter, however, Bell examines a number of images produced locally within a commercial studio in Samoa, which tend to disrupt our expectations of stereotypical tropical scenes and portraits. These photographs suggest more complex and nuanced viewings of Samoa during the 1890s, drawing attention to the fractured, unstable quality of the sites and spaces of the colony. This perspective on shifting and uncertain boundaries is complemented by Rod Edmond's emphasis on the complex and entangled geographies of so-called tropical diseases during the same period, at both the global and the metropolitan scales. The new science of tropical medicine is understood as a fraught project, seeking to protect European people and European space from the degenerative effects of tropicality, both within the tropics and within the temperate zone. Leprosy is treated here as the exemplary disease of tropicality not because it is in any way limited to the tropics but precisely because it its uncomfortable positioning within the discourse of "tropical medicine" that was designed to contain it.

VOYAGES

"On the Spot": Traveling Artists and the Iconographic Inventory of the World, 1769–1859

CLAUDIO GREPPI

Mr. Marten[s], a pupil of C. Fielding and excellent landscape drawer, has joined us. He is a pleasant person, and like all birds of that class, full up to the mouth with enthusiasm.

Charles Darwin to Caroline Darwin, Montevideo, 13 November 1833

"On the spot" is a phrase, rendered in French as *sur le lieu* or *sur le motif,* that conveys the spirit of a mode of landscape representation in which true knowledge of the natural world—and its botanical, zoological, human, and aesthetic forms—is based on direct observation in the field.[1] It became increasingly common in illustrated narratives of travel published in Europe, especially in England, from the late eighteenth century onward, signaling the significance that in situ observation had acquired for both the composition of images of landscape and the construction of authoritative knowledge about distant places. In what ways does this emerging discourse of field observation privilege visual images made "directly" from nature? How is it related to the emergence of a new figure—that of the "traveling artist"? And how does it both reflect and contribute to the construction of an iconographic inventory of the world, including its tropical landscapes?

Building on Bernard Smith's account of the impact of the Pacific voyages on European views of landscape, I will consider here the work of a succession of traveling artists, from William Hodges, who accompanied James Cook on his second voyage and subsequently worked in India, to Thomas Ender, who traveled extensively in South America. This body of work had a significant impact on European visions of the tropics, mediated as it was through the figure of Alexander von Humboldt, who was inspired by Hodges's representations of tropical nature. Humboldt's keen reflections on landscape painting and the aesthetics of landscape observation were in turn appropriated by a new generation of traveling artists, just as his observations on tropical landscape inspired naturalists such as

Charles Darwin. The result was a way of seeing, and knowing, in which the tradition of landscape art was fused with a new spirit of observation informed by the experience of voyaging around the world in the company of naval surveyors, meteorologists, and astronomers. This emergent epistemology of landscape is also evident in contemporary views and visions of European landscape itself.[2]

LANDSCAPES FROM OTHER HEMISPHERES

In 1786, William Hodges published a set of aquatints entitled *Select Views of India Drawn on the Spot,* the fruits of a journey along the valley of the Ganges between 1780 and 1783.[3] Hodges (1744–97) was already well-known as the painter who accompanied Cook on his second Pacific voyage. In 1777, engravings made after his sketches had been published in the official account of the *Resolution*'s voyage to the South Seas, together with the captain's journal and an atlas of charts and topographical sketches.[4] The task of making the engravings had been assigned to the "most eminent Masters," notably William Woollett, John Keyse Sherwin, and John Hall. Yet, for all their skill, the results could not compare to the richness of the atmosphere and the effects of light and shade in Hodges's originals. Alexander von Humboldt was later to acknowledge both the value and the limits of the technique of engraving evident in such illustrated accounts of voyages and travels: as he was to write in *Cosmos,* "even in the present imperfect condition of pictorial delineations of landscapes, the engravings which accompany, and too often disfigure, our books of travels, have, however, contributed considerably toward a knowledge of the physiognomy of distant regions, to the taste for voyages in the tropical zones, and to a more active study of nature."[5] Rather more to Humboldt's own taste, especially given the influence of the German natural philosopher Georg Forster, would have been the series of large-scale oil canvases that Hodges himself had painted after his return to England from the Pacific, prior to his departure for India.

Hodges's painted views were composed according to aesthetic conventions then dominant in Europe. In Forster's own journal of the *Resolution* voyage, we find a measured critique of Hodges's work and the engravings made after his drawings, in the course of an account of arrival at Middleburgh, in the "Friendly Islands":

> Mr Hodges designed this memorable interview in an elegant picture, which has been engraved for captain Cook's account of his voyage. The same candour with which I have made it a rule to commend the performances of this ingenious artist, whenever they are characteristic of the objects, which he meant to represent, obliges me to mention, that this piece,

in which the execution of Mr Sherwin cannot be too much admired, does not convey any adequate idea of the natives of Eaoowhe or of Tonga Taboo. The plates that ornament the history of captain Cook's former voyage, have been justly criticised, because they exhibited to our eyes the pleasing forms of antique figures and draperies, instead of those Indians of which we wished to form some idea. But it is also greatly to be feared, that Mr. Hodges has lost the sketches and drawings which he made from nature in the course of the voyage, and supplied the deficiency in this case, from his own elegant ideas. The connoisseur will find Greek contours and features in this picture, which have never existed in the South Sea.[6]

There was further criticism of Hodges's artistry by another acute observer, François de La Pérouse. As he wrote on 9 April 1786 (in the journal that Barthélemy de Lesseps would later bring back to Europe), "The drawings of these monuments [in Easter Island], executed by Mr. Hodges, give us but a very imperfect idea of what we are seeing." Looking at the engraved views, as well as the oils by Hodges, one can readily appreciate why such critics complained that the original sketch "on the spot" had been forgotten. Other canvases from the period between Hodges's travels in the Pacific and in India reflect the classical tastes that he had developed at the Royal Academy, under the instruction of Richard Wilson: the famous view of Oaitepeha Bay, Tahiti, also titled (significantly) "Tahiti Revisited," is one of the finest examples of these works (plate 1). This has been described as an example of "autofalsification," a readily understandable strategy given the cultural climate in which Hodges had been formed and—more particularly—to which he had to adjust in order to make his work marketable.[7] More surprising are the trenchant criticisms of Forster, reflecting a debate that stretched back at least to the publication of the results (both scientific and iconographic) of Cook's first Pacific voyage, from 1768 to 1771. In order to appreciate the sheer novelty of a painter like Hodges being on board the *Resolution,* we must move further back in time.

The team of specialists recruited for the *Endeavour* voyage included two expert draftsmen, Alexander Buchan (?–1769) and Sydney Parkinson (1745–71), who were to be employed in the service of Joseph Banks rather than Cook. Both lost their lives during the voyage, the former in Tahiti in April 1769, the latter in Batavia in January 1771. Buchan leaves relatively few traces in the historical record, apart from the fact of his death: his surviving drawings relate to the first stage of the voyage around Tierra del Fuego, including some significant images depicting the Fuegians' way of life. This work provides Bernard Smith with a telling case study of the abyss that divides the original sketch and the printed image, which appeared in a compilation of the accounts of John Byron, Samuel

Wallis, and Philip Carteret under the title of *An Account of the Voyages for Making Discoveries in the Southern Hemisphere,* edited by John Hawkesworth and published in London in 1773.[8] (A French edition was published in Paris the following year.) As John Beaglehole and others have noted, the task of the "professional" writer on voyages and travels in this period was to "polish the style of His Majesty's captains," and in this case the latter had no control over the "corrected" text. So, too, in the case of the scientific draftsmen, Buchan and Parkinson, whose views were revised by professional engravers—Francesco Bartolozzi and Giovan Battista Cipriani, originally from Florence and well known in London for their reproductions of Renaissance works. In their hands, Buchan's rough Fuegian hut was transformed into an Arcadian paradise, complete with cherubs.

Sydney Parkinson was first and foremost a landscape artist, with a particular penchant for drawing plants. His careful rendering of the vegetation of the Polynesian islands was transformed in Hawkesworth's edition into mere scenery, in which the highly particular physiognomy and associations of tropical plants were unrecognizable. (The commercial success of this edition, as well as such shortcomings, may account for the efforts of Parkinson's brother to publish at his own expense the travel journal in an edition "embellished with Views, and Designs, delineated by the author.")[9] Of more lasting significance were Parkinson's landscape drawings on the *Endeavour*—notably, coastal profiles and topographical sketches, which were traditionally the responsibility of maritime surveyors, trained in naval schools and academies. Parkinson, however, learned in the course of the voyage how to document the outline of coasts with the same skill used to render the physiognomy of plants. In the voyage of the *Resolution,* a similar process was at work: the expedition artist Hodges, this time working under Cook's orders, left his mark on a larger corpus of topographic and graphic images. In a process of reciprocal influence, the painter's own observations and drawings were in turn influenced by the maritime surveyor's way of seeing. At the same time, the naturalist Georg Forster was also drawn to experimenting with depictions of the profiles and atmospheric effects of icebergs during the long passage through the southern oceans.

On 11 May 1773, while at Dusky Bay (New Zealand), Cook made his own observations on the work of the *Resolution*'s artist: "Mr. Hodges has drawn a very accurate view of the North and of the South entrances, as well as of the other parts of the bay, and in these drawings he has represented the mood of the country with such a skill, that they will, without any doubt, give a much better idea than it is possible with words." From then on, the task of depicting "everything that it is impossible to describe" would be left to the expedition artist.[10] What particularly distinguishes the drawings made by Hodges during this voyage from traditional modes of landscape depiction is their treatment of the effects of light, clouds, and other meteorological phenomena. It would be far too simple to attribute this interest in atmosphere—which develops from the sketch, the water-color, and the oil painting, followed also by the engraving in aquatint—to a mere picturesque sensibility. This would be to ignore the fact that for three years Hodges had been working in close contact not only with naturalists and naval officials but also with the astronomer William Wales, who studied so-called meteors within the laboratory that was the South Seas. This provided him with the opportunity to render the physiognomy of celestial phenomena, just as Parkinson had studied the physiognomy of plants under the guidance of Banks and Daniel Solander.

Departing from his training as a landscape painter in the neoclassical tradition, Hodges had found in the Pacific a new interest in topography and meteorology that would profoundly modify the character of landscape painting itself, as well as influence the work of artists traveling to distant places. However, while it was heralded in the 1770s, this shift was yet to make its mark; and Hodges's subsequent fortunes were distinctly mixed. Having completed the drawings of his Pacific voyage and exhibited his canvases in London, he set out to make a new journey on his own, this time to India (at a moment of considerable difficulty, in the wake of wars and epidemics). While his Indian works had a certain commercial success (*Select Views* was reprinted in 1794, with a description of Indian architecture and an account of his travels), Hodges himself appeared increasingly to be an artist out of place, overtaken by political events in Europe. His attempts to reinvent himself as a painter of heroic and patriotic scenes resulted in repeated failure, and some commentators attribute his death in 1797 to suicide.

Hodges's death was preceded in 1793 by that of John Webber (1750–93), the artist on Cook's third voyage, and in 1794 by that of Georg Forster, in Paris. Cook himself had perished in the Sandwich Islands in 1779. The images that the Southern Hemisphere then transmitted to European culture were those of the death of Captain Cook, notably in the celebrated painting by Webber, and the disappearance of La Pérouse's own expedition in the Pacific. Might we identify in this period, then, something of a hiatus in the construction of an iconographic inventory of the world? Certainly, the role of the painter on grand voyages of circumnavigation had changed. John Webber, an artist of Swiss origin who had studied in Bern and Paris, had been chosen by Cook for his third voyage in 1776 "for the express purpose of supplying the unavoidable imperfections of written accounts, by enabling us to preserve, and to bring home, such drawings of the most memorable scenes of our transactions."[11] A new and significant task, no doubt, included the making of topographical and scientific sketches in which Webber proved to be particularly prolific. However, the most memorable "scene" of the whole voyage would be the death of the captain as depicted in Webber's painting, reproduced in a variety of contexts (including a theatrical *pièce* in four acts performed in Paris). In addition to the publication of engravings based on his drawings with the journal of the expedition in 1784, edited by James King, Webber published sixteen plates of *Views of the South Seas* between 1788 and 1792; these latter would eventually become better known in the color aquatints published posthumously by Boydell in 1808.[12] The more strictly nautical and scientific aspects of drawing then reverted to naval officers and midshipmen, whose training was adapted in order to aspire to the standards achieved on Cook's voyages. Thus when George Vancouver—who accompanied Cook on his third voyage—returned to Pacific waters on his own expedition (1790–95), there was not a single painter on board.[13] Artists were only to be employed when all the graphic materials were collected in London, in order to prepare the reproductions for publication.

While La Pérouse's final voyage ended in disaster, the sketches of the expedition artist, Gaspard Duché de Vancy (?–1788), were brought safely to Paris by Barthélemy de Lesseps. The artist, who worked on the *Boussole,* had a role similar to Webber's, while the task of making landscape drawings and topographical sketches was assigned to a certain Blondela, an official on the *Astrolabe.* Even when he was given the job of correcting Hodges's views in the name of accuracy (as in the case of the Easter Island images mentioned above), the artist's drawings were designed above all to testify to the French presence on the island, amid scenes of mysterious monuments, rather than to document the physiognomy of a profoundly distinctive landscape. Rather than reproducing aspects of nature, Duché de Vancy's drawings were intended to illustrate the text. It is as if everything had already been represented: in Kamchatka, the French voyagers were already familiar with at least the appearance of the various personalities of the Russian colony, as published in *A Voyage to the Pacific Ocean* (1784).[14] What was left to de Vancy was the depiction of scenes of local life or encounters with the inhabitants, in which the noble savage appears increasingly more ignoble.

Had La Pérouse's voyage met with a different fate, the whole of the graphic material—together with the objects and curiosities collected during the expedition—would have been under the captain's control. Was, then, the freedom given to Hodges (as well as to Forster and Wales) something of an exception to the rule? Other voyages during the late eighteenth century provide further examples. Thus, in the course of Nicolas

Baudin's expedition, a conflict arose between the artists and the captain, causing Jacques-Gérard Milbert and other artists to disembark in Mauritius in 1800.[15] The task of producing scientific drawings was then assigned to Charles-Alexandre Lesueur (1778–1846), who had originally embarked as helmsman but who would later become one of the most skilled zoological draftsman under the instruction of the naturalist François Péron. Before proceeding further, however, with the subsequent history of image making in maritime expeditions (especially after 1815), I want to return to India in order to consider other aspects of the iconography of tropical landscape at the end of the eighteenth century.

British artists traveling in India were more or less dependent on the East India Company. Thomas Daniell (1749–1840), only five years younger than Hodges, presented himself to the company not as a painter but as an engraver wishing to open an aquatint workshop in Calcutta. On these terms, he managed to obtain permission to work in India both for himself and his nephew, William Daniell (1769–1837), who was only fifteen years old when they left England in 1784. Hodges had returned from India the year before, having been inspired by the light of the subcontinent as much as by that of the South Seas. Nonetheless the "impressionist" style of his images of landscape and monuments failed to satisfy Thomas Daniell, who was seeking to develop an ultrarealist iconography, though in accordance with the Claudian model of landscape as far as composition is concerned.

The Daniells spent four years working in Calcutta, working on views of the new British capital that had a market among the colonial elite before they were appreciated in the mother country. From 1788 to 1791, they undertook a lengthy journey into the interior, initially following Hodges's itinerary, which provided an opportunity to redraw some of his views in the mode of ultrarealism. The systematic use of the camera obscura informed their "correction" of the hazy impression that the Indian atmosphere had imprinted on the images of their predecessor. The Daniells' curiosity drove them further on, up to Srinagar and the foothills of the Himalayas, where their expedition (accompanied by a small escort) reached places beyond the control of the company. The traveling artists thus enjoyed a degree of political as well as financial autonomy, and their itinerary was shaped less by the demands of the colonial authorities in Calcutta than the developing scientific interests reflected in the foundation of the Asiatic Society in 1784. The relationship between the Daniells' Indian journey and the world of science is a subject hinted at by their biographer Mildred Archer, and it deserves further attention (see also chap. 8).[16] They set themselves the task of visualizing unfamiliar places across an immense continent, and their technical skill in both drawing and engraving allowed them to transmit the character of their observations to a wide audience. Moreover, the fact that their iconographic inventory was being constructed according to an explicit reappropriation of Claude's model indicates that the academic culture of landscape could enable the traveling artist to enlarge the field of knowledge rather than simply representing a limit to expressive innovation.

The last years of the Daniells' extended stay in India were dedicated to a new journey of pictorial exploration, this time to the south, and finally around Bombay. This was again exploration in more than a metaphorical sense, for parts of Madras had not yet been occupied by company forces. At every city they visited, the artists' itinerary crossed those of other travelers who shared their aesthetic and scientific interests. In Calcutta at that time, the best-known painter was the Anglo-German Johann (John) Zoffany (1733–1818), a personal friend of Joseph Banks, who specialized in portraits, having like Banks failed in his bid to accompany Cook's second expedition. In Madras and Bombay, the Daniells encountered painters who were working on iconographic surveys of landscapes and monuments. In Madras, for example, they accompanied the artist Robert Home (1752–1834) on some of his excursions around the city and came to appreciate his sketches of the surrounding area.[17] In Bombay, their interlocutor was John Wales, who entrusted to them his sketches at the mountain of Ellora; following the artist's death in 1795, the Daniells published engravings after his drawings, incorporating within their iconographic repertoire a place they were unable to see first hand.[18] The reproduction of original drawings was in this case overseen by the returning traveling artists, who were evidently more able to interpret the particular character of views of distant places than professional engravers.

Having spent nearly ten years in India, Thomas and William Daniell returned home, devoting themselves principally to reproducing views made by themselves and other travelers, including *Antiquities of India* (fig. 2.1).[19] In 1795, they began publication of a series of Indian views under the title *Oriental Scenery,* a plate every month, in association with Longman, Hurst, Rees, and Orme.[20] While some volumes in the series—each usually containing twenty-four plates—did not achieve the anticipated success, mainly because of their high cost, the market for oriental iconography had at least been established. Subsequently, between 1812 and 1816, a more affordable edition was issued, in quarto, which contributed to the extraordinary popularity of the Daniells' images of India. Their work was also reproduced in tapestry and porcelain, while its analytical precision also influenced the architectural pastiches then in vogue in Georgian England. Rather than treating the Daniells' body of work simply as a reflection of exotic and oriental tastes and crazes, however, we should consider their

FIGURE 2.1 *The Entrance to the Elephanta Cave,* plate 7, from Thomas Daniell and William Daniell, *Antiquities of India: Twelve Views* (London, 1799) (by permission of the British Library)

function as intermediaries between the new generation of artists working on the spot and the metropolitan centers of expertise, both experimenting with and diffusing new views of distant places.

During the closing decades of the eighteenth century, when the archives of natural history were enriched with a wealth of new specimens, the oriental and southern hemispheres were more fully incorporated into the European imaginary. And in the same period, the representation of landscape took a particular scientific turn.[21] Alexander von Humboldt, who would become the protagonist of this new scientific interest in landscape, was determined to travel south, though it was not clear at first whether he would go to India or Africa. His ultimate choice of destination, equatorial America, was determined by various contingencies, including the political situation in Europe and the support of the Spanish court, enabling him to travel through the tropics with a degree of autonomy.

WITHIN EUROPE

In *European Vision and the South Pacific,* Bernard Smith highlights the innovative impact that the work of traveling artists had back in Europe, challenging the common assumption that they were invariably dependent on conventional models of observation and representation. In this context, references to the picturesque, to the sublime, or to exotic taste are so ubiquitous that they do not in themselves explain anything. Indeed, it would be worth considering the extent to which contemporaries— Georg Forster, for example—criticized the effects of classical models on the reworking of images made on the spot. Moreover, as we have seen in the case of the Daniells, the use of a Claudian framework did not necessarily undermine the project of delineating the physiognomy of Indian landscapes. The contrasting methods of Hodges, in the same sites, throw further light (literally) on the issue: rather than reflecting a different pictorial language, they stemmed from experiences on his Pacific voyage. It is not so much a matter of European convention producing the iconography of distant places but, rather, of the experience of travel modifying the ways in which even the European landscape is represented.

What is at stake here is not simply the technique of representation but also the quality of the object represented as landscape. According to Martin Kemp, the major paradox in the history of spatial representation in Europe between the eighteenth and nineteenth centuries is that "the theoretical aspects of optical and geometrical space became ever more widely discussed in the literature on the arts and sciences at the same time as the hold of perspectival techniques on the practice of the 'Fine Arts' was being radically loosened."[22] Gradually limited to the métier of specialists in descriptive geometry, perspective became less and less important to the practitioners of academic art. At the same time, topographic painting— hitherto regarded as a lowly art by academic critics—could be redeemed by situating it within a major new project: that of assembling an iconographic inventory of the landscapes of the earth.

"Topography, of course"—writes Bernard Smith in the context of Hodges's paintings of Tahiti—"had always been given a humble place at the bottom of the academic table, but here was an attempt to elevate exotic topography to the high places reserved for the ideal landscapes of Claude, the heroic landscapes of Poussin, and the picturesque landscapes of Salvator Rosa."[23] Smith's argument, which draws attention to Hodges's rendering of "tropical atmosphere" blending the documentary and the ideal, also needs to be seen in the context of renewed academic interest in the description of places not so exotic. For example, among Goethe's papers, we read the following in a letter "on landscape painting" by Jacob Philippe Hackert (1737–1807): "What I desire is that a botanist recognises at once trees, plants and other leaves on the foreground: this is a good way of reproducing nature without mannerisms."[24] The author of this letter, sent from Florence in 1806, was the official artist at the court of Ferdinand IV in Naples from 1786 to 1800. It was in Naples, in 1770, that Hackert had met William Hamilton, the British ambassador interested in the study (and representation) of volcanic phenomena, together with the Neapolitan painter Pietro Fabris. It was also in Naples seventeen years later that Hackert met Wolfgang Goethe, to whom he gave drawing lessons in the weeks before they departed to Sicily in the company of the young painter Christoph Heinrich Kniep.

Far from disowning the topographical view, Hackert transformed it into a genre worth exhibiting at the royal galleries of Caserta, where a series of large canvases depicting maritime views of the kingdom were on display. Hackert's work also included the production of engravings (etchings, in particular) and teaching: his interest in a "nonmannerist" treatment of vegetation, as expressed above, was reflected in his album of trees, a work effectively designed to be a manual of landscape art. While he did not publish his own volumes of sketches and watercolors, his influence may be detected in the *Campi Phlegraei,* a series of views published by the British ambassador himself in 1776.[25] This marks the beginning of a long sequence of illustrated books through which it is possible to trace—more readily than with any other material—the genealogy of images of places in Europe at the end of the eighteenth century. But what kinds of places? Initially, the focus of interest was on volcanoes and mountains as both subjects for visual representation and sites of scientific knowledge.

A pioneer study of the Swiss Alps was the work by Gottlieb Siegmund Grüner, *Die Eisbirge des Schweitzerland,* in which the glaciers were depicted after the drawings by Samuel Hieronymus Grimm (1733–94). Published initially in Bern in 1760, it was the Parisian edition of 1770 that really aroused curiosity about the alpine world. Six years later, a work entitled *Vues remarquables des montagnes de la Suisse* attributed to Rodolphe Hentzi was published in Bern, accompanied by aquatints after drawings by Caspar Wolf (1735–98), a German-trained Swiss painter.[26] Further collections of images of the Swiss Alps were published under the same title by Wolf himself (Bern, 1778; Paris, 1787–91) and by Charles-Melchior Decourtis (Amsterdam, 1785).[27] The expression *vues remarquables* seems to have been universally employed only with reference to the Swiss Alps: it is not used for example in the work of the famous propagandist of the glaciers of the Savoy, Marc-Théodore Bourrit (1735–1810), who published his *Nouvelles description des vallés de glace* in 1783, the same year in which the naturalist Horace-Benedict de Saussure was preparing his expedition to Mont Blanc.[28] In the introduction to his *Voyages sur les Alpes,* de Saussure acknowledged his debt to Bourrit's draftsmanship with the following words: "The views of the mountains depicted here were drawn on the spot by M. Bourrit with an exactitude that one could say 'mathematic,' for often I have verified the proportions with a 'graphometer' and no mistake was found. He has even sacrificed part of the effect of his drawings on behalf of this exactitude, most notably in experimenting with the detail of the inclinations and in highlighting the contours of the rocks."[29] De Saussure proceeded to lament that the technique of stipple used in engraving was not capable of rendering the visual impact of glaciers as well as Bourrit had achieved in his drawings.

In the early illustrated volumes of travels, the producer of the original sketches had a secondary role, though it became increasingly common for the name of the artist to appear on the frontispiece. However, while the principal artist might be acknowledged, that of the assistant draftsman often continued to be subordinate. Consider, for example, the case of poor Kniep, Goethe's young companion in Sicily, who submitted to the most humble tasks, carrying the luggage or helping his mentor get over his seasickness. His own drawings remained, by contract, the property of Goethe; indeed, we very rarely find any reference at all to Kniep in the many editions of *Italienische Reise*. While artists such as Hackert were Goethe's social equals, Kniep was clearly his servant and willingly submitted himself to all sorts of discomforts, climbing hills and trees in order to obtain the best viewpoints for the "master": though Goethe appreciated his rapidity and precision, he regarded his contribution as that of providing "mementoes." In this context, a little anecdote reveals the extent to which Goethe, like Forster, was well aware of the possibility of falsifying the landscape view: "Kniep made one sketch of an interesting distant view, but the middle and foreground were so awful that he introduced into the latter an elegant group in the style of Poussin that cost him little trouble and transformed the drawing into a delightful little picture. I wonder how many 'Travels of a Painter' contain such half-truths."[30] Elsewhere, Goethe compared his own impressions of Monte Pellegrino with the reproductions he remembered from the *Voyage pittoresque de la Sicile,* published in 1781–86 by the abbot de Saint-Non, judging the latter to be "inaccurate."[31] (The same term was used by La Pérouse in his criticism of Hodges.)

Goethe had been particularly enthusiastic about the prospect of his journey to Sicily: "Given my temperament, this trip is salutary and even necessary. To me Sicily implies Asia and Africa, and it will mean more than a little to me to stand at the miraculous centre upon which so many radii of world history converge."[32] The artist who had illustrated Saint-Non's *Voyage pittoresque,* though his name did not appear anywhere in the title of the work (nor even in recent reproductions), was Claude-Louis Chatelet (1753–94). Another work on Sicily that Goethe might have been familiar with was a set of 264 aquatint plates published under a nearly identical title by Jean-Pierre Houel (1735–1810) in 1782.[33] Houel had first visited Naples in 1770–71, and he returned to the region in 1776 to visit Sicily, Malta, and the Eolies. His own *Voyage pittoresque* not only exploited his extraordinary skills as a scientific draftsman but also combined graphic image with literary narrative in a publishing formula that was beginning to be successful in France in the 1780s.

In addition to the illustrated works by Saint-Non and Houel on Sicily, the year 1782 saw the publication of an eight-volume *Voyage pittoresque de la France* by Jean-Benjamin de Laborde (1734–94), a courtier of Louis XV, musician and publisher, who would end up on the guillotine as did Chatelet.[34] Laborde was also engaged in other demanding editorial enterprises, including the *Tableaux topographiques, pittoresques, physiques, historiques, moraux, politiques, littéraires de la Suisse et de l'Italie* (1780–86), published in four volumes in folio and fourteen in quarto, which included 1,200 illustrations produced by an army of draftsmen (including Houel),[35] and the *Description générale et particulière de la France* (1784–88) published in twelve volumes. The formula of the *Voyage pittoresque,* consisting of descriptions of places based on a combination of text and images, would continue to characterize French travel works. In London during the same period, publications tended to be divided between written narratives of travels, occasionally accompanied by related iconographic materials, and series of free-standing vistas such as those by the Daniells, under the less ambitious title of *Voyages* or *Select Views.*

Other examples of the literature of "picturesque travels" included works by Auguste-Florent de Choiseul-Gouffier on Greece (1782–1809, with drawings by Jean-Baptiste Hilaire and Louis-François Cassas), by Cassas on Syria, Palestine, and Egypt (1798–99, with texts by various authors, including Constantin-François Volney) and on Istria and Dalmatia (1802), by Louis-Joseph de Laborde on Spain (1806–18), and by Jacques-Gérard Milbert on the island of Mauritius and the Cape (1812).[36] Also with a similar title, and using the same template, there was a description of North Cape by Anders Skjöldebrand (Stockholm, 1801) and, finally, the *Viaggio pittorico della Toscana* edited by the abbot Francesco Fontani, with engravings by Antonio Terreni, published in three volumes (Florence, 1801–3).[37] The structure of the latter, whose text deserves to be studied as much as the images, confirms the extent to which the notion of an iconographic inventory had been enlarged in the second half of the eighteenth century, extending from urban vistas and principal monuments to an ever-widening territory, eventually embracing those areas of Tuscany, such as Lunigiana and Maremma, situated well off the beaten track.

What is significant here is that even such a well-known and celebrated region as Tuscany could be the subject of the new spirit of observation. Fontani followed the French model, understandably translating *pittoresque* into *pittorico;* yet in the views that punctuate the work, there were no concessions to the picturesque aesthetic, either in the choice of subject or in the style of pictorial representation. Equally, the term *viaggio* in the title of such publications stretches a point: the voyage itself is a pretext, a reconstruction in the cabinet of a possible suggestive itinerary between places and scenes whose significance depends on a long process of the sedimentation of knowledge rather than the fleeting impressions of a traveler.

RETURN TO THE TROPICS: IN HUMBOLDT'S WAKE

The formula of the picturesque voyage was later extended by the artist Jean-Baptiste Debret (1768–1848) beyond Europe, in his *Voyage pittoresque et historique au Brésil* published in Paris in 1834, with lithographs by Charles Etiènne Pierre Motte and the brothers Thierry.[38] The work consists of a succession of views made during a sixteen-year stay in Rio de Janeiro. Debret had initially traveled to Brazil as a member of the French artistic mission commissioned by Don João VI, and while in Rio he had the opportunity to attend and depict the coronation of Pedro I. His reputation in Paris had been established by his classical depictions of episodes in the life of the emperor in a Roman manner. In Brazil, he became an attentive observer of the tropical forest, initially depicted through an inventory of natural forms and later as the setting for the rituals of the indigenous Tupis. As with Houel's *Voyage pittoresque,* Debret was also the author of the accompanying text, which consisted of general introductions to the three volumes and captions for each plate, as well as hints for travelers intending to visit the region and to study its inhabitants. Debret used his acknowledgments to recall the influence of Alexander von Humboldt on his decision to travel to Brazil in 1815.

Humboldt's interest in the work of artists is reflected in much of his writing during this period. His *Essai sur la géographie des plantes* (1807) and *Ansichten der Natur* (1808) include extended discussions of the necessity of adequate graphic representations of scientific observations of nature. And in his *Vues des Cordillères et monumens des peuples indigènes* (1810), Humboldt reproduced—with the assistance of Parisian specialists in aquatint, lithography, and engraving—his own sketches made during the Andean and Mexican stages of his journey.[39] Pierre Antoine Marchais and Wilhelm Friedrich Gmelin produced nearly all drawings for this work, while Friedrich Arnold and Louis Bouquet were responsible for the diagrams: the final results thus represent the collaborative labor of many hands. The plates convey a somewhat cold feeling, lacking the light and atmosphere that Humboldt had so admired in Hodges's paintings and that he described with such passion in *Ansichten der Natur.* In its combination of text and images, the structure of the *Vues des Cordillères* was similar to the model of the *voyages pittoresques*. However, instead of a hypothetical voyage, we are presented with a hypothetical juxtaposition of forms of landscape and the cultural expressions of pre-Columbian civilization. To judge from the character of the images and the reproductions of Aztec codices from European archives, the enterprise reflected an essentially eighteenth-century idea. Humboldt may well have been dissatisfied with the results, given that he lost interest in further publications of this work, in contrast to the *Ansichten,* which was extended and reprinted in two consecutive editions, in 1829 and 1849. In the latter, Humboldt devoted a long footnote (extending over five pages) to an essay already published in *Cosmos* entitled "Landscape Painting, in Its Influence on the Study of Nature," in which he considered improvements in techniques of pictorial representation that had arisen alongside the extension of geographical knowledge. The elderly Humboldt was very much attached to this text, in which the literary "aspects" of nature were enriched by technical as well as historical reflections on the translation of observations into images and on the role of landscape painters, including great masters of the past, including Ruysdael, Poussin, and Claude—further evidence of what he called "the ancient link between science, art and poetry." In his account of more contemporary works, Humboldt was decidedly critical of the choice of artists to accompany transoceanic voyages: "The end of the voyage may thus have drawn near before even the most talented among them, by a prolonged sojourn among grand scenes of nature, and by frequent attempts to imitate what they saw, had more than begun to acquire a certain technical mastery of their art." [40]

While Humboldt did not refer explicitly here to any particular artist, the names cited elsewhere in his essay provide some indication of those he regarded as the "most talented"—including both those who influenced Humboldt and those who had been inspired by his works. Among the former, Hodges was mentioned alongside Ferdinand Bauer (1760–1826), an Austrian scientific draftsman invited by Joseph Banks to London in 1798. Bauer's association with Banks had begun in Vienna (in connection with Nikolaus von Jacquin), but was to develop still further in his work with the botanist Robert Brown on the 1801 *Investigator* expedition to the coasts of New Holland under the command of Matthew Flinders. This was a voyage that proved the exception to Humboldt's rule, as its principal objective was to survey the coastlines of the fifth continent, henceforth to be known as Australia. The focus of the expedition shifted gradually from the sea to the interior of the new colony, to its fauna and flora and, especially, its landscapes. Bauer's drawings were to be become well-known through the published work of Brown (in which images of the gum tree and the

platypus appeared for the first time in Europe): indeed, even Goethe referred enthusiastically to them.[41]

Of equal interest to Humboldt would have been the drawings by the
landscape artist William Westall (1781–1850), who composed numerous
coastal views, geological profiles, and botanical sketches in the course of
the same expedition. Flinders himself considered Westall's coastal profiles
to be the best ever made in the Pacific.[42] Westall's biography bears some
similarities to that of the Daniells, not simply because he was in fact
William's brother-in-law and had been instructed in the same school but
also because after the Flinders voyage he remained in the Orient search-
ing for new scenery. In 1804 he requested permission from the East India
Company in Canton to venture into those parts of India that "have hitherto been but little visited of artists."[43] While in Bombay he visited the
Mahratta Mountains and made sketches of the cave temples, and he subsequently traveled to Jamaica. Having returned to London, he continued
(like the Daniells) to work on the reproduction of vistas, from sketches
made by himself and other travelers, for specialized publishers including
Ackermann and Murray, even contributing to the illustrations for the narrative of Parry's Arctic voyage in 1821.[44] While in some respects his experience was similar to William Daniell's, his work was also definitely shaped
by his voyage in the South Seas accompanied by naval officers and naturalists, just as Hodges's had been. Several of his works, especially his watercolors, effectively synthesize the two modes of expressing Indian landscape that were competing for primacy at the turn of the nineteenth
century: the topographical precision of the Daniells, joined with the lu-
minosity of Hodges.

Notwithstanding Humboldt's strictures on the quality of artistic depictions of landscape scenery, the collaborative work of artists and naturalists
on board sailing vessels, as well as on land, had yielded much fruit and
would continue to do so after 1815. One key factor was the capacity of
the European scientific community, and of Humboldt in particular, to
guide the work of traveling artists at a distance. Jean-Baptiste Debret, who
went to Brazil in 1815 at Humboldt's suggestion, is a case in point. Another example is provided by the Russian artist Ludovik (Louis) Choris
(1795–1828), a member of Otto von Kotzebue's expedition to the South
Seas and the Bering Strait, who declared (in the same year) that "this enterprise has been inspired by the immortal work of Baron von Humboldt,
the *Tableaux de la Nature*."[45] Before the official account of the voyage undertaken by the Russian-German navy had appeared, Choris published his
images in *Voyage pittoresque autour de monde* (Paris, 1820), in collaboration
with the poet Adelbert von Chamisso (1781–1838), who was also responsible for many of the drawings of plants and landscapes.[46] In the context

of the French publishing tradition discussed above, the term *pittoresque* had now acquired a completely new sense, much closer in fact to the idea of the "aspects" (*Ansichten*) of nature. A later work by Choris carried an explicitly Humboldtian title, *Vues et paysages des régions équinoctiales* (1826).[47] Humboldt was also the inspiration for Choris's own voyage in 1827 to the equatorial regions of the new continent: soon after his arrival, however, he met a violent death in Vera Cruz.

What Choris failed to achieve would be accomplished over the following decades by a series of German-trained artists who were to traverse and describe the continent of Latin America, many of them guided at a distance by the aging Humboldt. In 1817, the Viennese painter Thomas Ender (1793 –1875) traveled to Bahia and São Paulo on board the *Austria,* which was carrying the archduchess Leopoldina to her marriage with Dom Pedro in Brazil. In just one year, Ender produced seven hundred works in drawings and watercolors. While there is no evidence of direct contact between Ender and Humboldt, it is worth noting that the zoologist Johann Baptiste von Spix and the botanist Carl Friedrich Philipp von Martius were on board the same ship, the latter becoming one of the most accomplished draftsmen of tropical flora. In the splendid plates of Martius's *Historia naturalis palmarum* (published between 1823 and 1850), plants were depicted in the context of their forms of association, as Humboldt had proposed in his *Géographie des plantes,* rather than in isolation (see chap. 3 below, esp. fig. 3.5).

In 1821, Johann Moritz Rugendas (1802 –58) was appointed draftsman to the Russian scientific expedition to Brazil led by Baron von Langsdorff. Descended from a well-established Augsburg family that included at least six generations of painters, he had learned to draw in the family workshop. On his return to Europe, Rugendas met Humboldt in Paris; after showing him his drawings, he was encouraged by the naturalist to return to the Americas. Over a period of sixteen years, Rugendas traveled through Mexico, Chile, Peru, Bolivia, Argentina, Uruguay, and Brazil. Collections of his drawings were subsequently published in several editions, in Berlin, Paris, and London, under the title *Malerische Reise.* (Plate 2 shows an example of his oil paintings of the Brazilian tropical forest.) Rugendas was to be followed to the Americas by Ferdinand Bellermann (1814 – 89) and Eduard Hildebrand (1818 – 69), both of whom had been trained in Berlin and received financial support for their travels from Frederick Wilheim IV, on Humboldt's recommendation. Bellerman worked in Venezuela for four years, from 1842; Hildebrand in Brazil and the United States, between 1844 and 1845. Their trips were shorter than Rugendas's, but more directed: that is to say, they were guided more directly by Humboldt who followed their journeys from his study in Berlin (in which the eighty-year-old naturalist would later be portrayed by Hildebrand sitting among his books, maps, and specimens).[48]

Bellerman followed in Humboldt's footsteps through Venezuela, composing images (of the forest, the *llanos,* the Cordillera) that his predecessor had described forty years earlier in words. With the naturalist confined to his study, his mobility restricted to short distances, it was the painter who could travel: and between them they established a collaborative relationship that might surpass the shared experience of traveling artists and naturalists on board ships. The Berliner Albert Berg (1825–84) thus followed Humboldt's route through Colombia and published, in Düsseldorf and London, his extraordinary plates entitled *Phisiognomie der tropischen Vegetation Süd-Americas* (1854), which represented—as the title indicates—a series of views (*Ansichten*) of the primeval forest (*Urwäldern*) along Magdalena River and in the Andean region of Nueva Granada. A letter from Humboldt himself provided the introduction (see also chap. 3, esp. fig. 3.3).

Other painters were attracted to the idea of retracing Humboldt's itinerary in order to redraw, in a new spirit and with new techniques, the scenes described in *Vues des Cordillères*. The American landscape artist Frederic Edwin Church (1826–1900), for example, traveled through the Andes between 1853 and 1857 in order to depict the scenery of Chimborazo and Cotopaxi in a number of impressive paintings exhibited in New York and London. The exhibition of his work, most notably the spectacular canvas entitled *The Heart of the Andes* (plate 3), gave Church the opportunity to travel to Berlin in order to meet the venerable Humboldt, but, alas, he arrived too late. The man who contributed so much to the rediscovery of the landscapes of the American tropics had died in 1859, at the age of ninety.

Far from being a mere scholarly diversion, then, I would suggest that Humboldt's essay on landscape painting was fundamental to his later researches, insofar as it testified to the long-distance relationship with places that the labor of painters made visible through their rendering of particular forms. As Humboldt makes clear, however, the scientific role of painters was enabled by the poetics of landscape that preceded it, and ultimately this could be traced back to the "great masters," though their work had been limited by European horizons. A degree of continuity between Claude's landscapes and the new topographical views is suggested not merely by explicit references to tradition but also by the academic training that artists received. If there was a clear tendency for traveling artists gradually to break free from the taste for classical landscape, this may be attributed precisely to the experience of voyaging to distant lands. Yet Humboldt's influence was not confined to those artists who traveled across the globe. Carl Gustav Carus (1789–1869), a surgeon and painter influenced by *Naturphilosophie,* used his *Briefe über Landschaftsmalerei* (1831) to argue the case for a "geognostical" vision of the morphology of the landscape, in parallel with Hackert's claims for botanical objectivity.[49] Indeed, some of the most influential theorists of nineteenth-century visual culture, including John Ruskin (in *Modern Painters*) and Jacob Burckhardt (in *Die Kultur der Renaissance*), may be found among Humboldt's most enthusiastic readers.

Humboldt's own career as an explorer came to an end with his excursion to the frontiers of China in 1829. While relatively few of his admirers encountered him in person, his works found pride of place in the libraries of traveling naturalists and artists—the most notable example being Charles Darwin, whose collection on board the *Beagle* in 1832 included Humboldt's *Relation historique* of his five-year expedition on the New Continent. For the first year of the *Beagle* voyage, up to Montevideo, the artist on board was Augustus Earle (1797–1838).[50] His work was not often reproduced in engravings, with the exception of a series of plates of New Zealand published in London in 1838, the year of his death.[51] His life as a traveling artist was restless, to say the least. Born in 1797 in the United States, he was trained (though not formally registered) at the Royal Academy in London. In 1821 he sailed to Brazil, where he produced a portrait of himself on the summit of the Corcovado Mountain above Rio de Janeiro, expressing through a gesture of astonishment a feeling of awe at the sight of bay (fig. 2.2). In another self-portrait, again alone (with his dog and gun), Earle pictured himself reclining on the rocky coastline of Tristan da Cunha, where he stayed for several months in 1824 in order to depict "a spot hitherto unvisited by an artist."[52] This phrase is almost identical to the one used twenty years earlier by Westall, in his quest for novel scenery and impressions in India (cited above): it is almost as if the world has become too small for painters anxious to represent original views "on the spot." By January 1825, Earle was in Tasmania (Van Diemen's Land), followed by Australia, where—in addition to earning his living by painting portraits of colonial personalities—he traveled into lesser-known regions such as the Illawarra, pioneering the development of colonial topography. In 1827 he visited New Zealand, and the following year he left Sydney, traveling to England via Guam, Manila, Madras, and Mauritius. His subsequent employment as the artist on the *Beagle* voyage was, as we have seen, interrupted at Montevideo: his place alongside the naturalist Charles Darwin was taken by Conrad Martens.

Conrad Martens (1801–78), with whom we conclude, brings us full circle to the South Seas and Australia. He had studied landscape painting in London under Copley Fielding, a master who later taught John Ruskin,

FIGURE 2.2 Augustus Earle, *View from the Summit of Corcovado Mountain,* ca. 1822, watercolor (by permission of the National Library of Australia, Canberra)

and who may—according to Bernard Smith—have been the source of Martens's original interest in the effects of weather. As a consequence of his training, Martens would also become a keen reader of Turner's *Liber studiorum.* The crucial moment in his formation, however, was on board the *Beagle,* with Humboldt as his guide through the medium of Darwin, with whom he developed a close working relationship. Apart from learning to observe geological and botanical phenomena, Martens would also have been influenced by the Humboldtian search for the elements that defined the mood of particular places. The idea of "mood" in this context is not very far from the Humboldtian conception of the physiognomy of nature. As Humboldt himself put it: "The azure of the sky, the form of the clouds, the vapory mist resting in the distance, the luxuriant development of plants, the beauty of the foliage, and the outline of the mountains, are the elements which determine the total impression produced by the aspect of any particular region. To apprehend these characteristics, and to reproduce them visibly, is the province of landscape painting."[53]

★ 3 ★

The Stimulations of Travel: Humboldt's Physiological Construction of the Tropics

MICHAEL DETTELBACH

CROSSING THE QUINDÍO PASS

At the end of September 1801, Alexander von Humboldt and his companion, the botanist Aimé Bonpland, were impatient to begin crossing the western ridge of the Andes via the Quindío Pass (in present-day Colombia). They were being forced to hire *silleros,* men who made scant livings carrying travelers, strapped to their backs in bamboo chairs, across the mountains (fig. 3.1). Neither their Creole hosts in Ibagué nor the *silleros* themselves would allow the two to cross on foot: the path across the Quindío massif was a narrow river of mud, often eroded several meters below the surrounding terrain, and filled with deep, leg-breaking ditches worn by the steps of the countless oxen that carried goods, not people, across the pass. Crossing the Quindío Pass demanded an enormous amount of skill and dexterity, and Humboldt admired the ingenuity and strength that the *silleros* applied to their trade. (Humboldt later wrote in his journal that crossing the Quindío Pass "was like walking on a ladder hung over an abyss at a 30° angle," twelve hours a day for two weeks straight). The bamboo chairs and leather suspension system developed to distribute the weight of their burdens were extremely clever. Where the path was overgrown with thorny bushes that tore at their feet, the *silleros* wore stilts. And yet falls were extremely rare.

But Humboldt deplored the inhumanity of "riding human beings" like horses. Some travelers prodded their mounts with canes and spurs, and *silleros* weakened by exertion or the miasmas of the trenchlike path were often left to die by the way. Not only was it barbaric; it was uneconomical. Humboldt decided that *andár en carguero* was a barbarity perpetuated by the enforced monopoly of the *silleros,* the "effeminacy" and

FIGURE 3.1 *Passage du Quindiu, dans la Cordillère des Andes,* from Alexander von Humboldt, *Vues des Cordillères et monumens des peuples indigènes de l'Amérique* (Paris: Gide, 1810)

"overrefinement" of Europeans, and the weakness of enlightened administration in the Spanish colonies—a perfect example of the caste prejudice that so thoroughly undermined colonial economy and society. Every sturdy, impoverished male in the district was drawn to the trade, their labor quickly and unproductively expended in exhaustion and illness. In this way, a whole class of industrious men was given over to vagabondage, exhaustion, and illness, impelled by their natural longing for freedom and independence. It behooved the colonial government or enlightened private persons to improve the road over Quindío and to eliminate or restrict the trade, "to give human energy a direction more useful to society."

So despite the obstacles before him, Humboldt resolved to traverse the pass on foot. The crossing (he wrote in his journal) would be "nothing that extraordinary for men like us, who customarily walk 6–9 leagues a day, who have waded through rivers and lived for months in the forests among the Indians." In doing so, Humboldt was defying his hosts with a demonstration not just of fellow feeling, but of a real sensitivity to physical effort, the historical struggle of human forces with a dynamic nature. In other words, Humboldt was not just manifesting fellow feeling in his sympathy for the *silleros,* a humanitarian sensibility; it was a political-economic sensibility, a vision of the struggle of physical forces and its effects on human societies—humanitarian or philanthropic in a larger, more philosophical sense.

Consequently, he was horrified to find himself sitting in a bamboo chair lashed to the back of a *sillero* so that the strapping young man could assess the weight of his prospective burden. In the end, Humboldt convinced the *silleros* to allow him and Bonpland to traverse the pass on foot by only demanding that he be given the chair and carrying one of them around the room. Humboldt took barely three steps before stumbling, but the piece of theater worked: the five *silleros* agreed to carry delicate barometer tubes, thermometers, and geodetic instruments and crates of mineralogical and botanical specimens and manuscripts rather than the two mad naturalists. Two weeks later, Humboldt and Bonpland arrived in Cartago, feverish, unshod (their boots having disintegrated in the mud), their feet sodden and shredded by rocks and thorns, but their physical and moral virtue intact, indeed, now visibly displayed.[1]

A number of themes come together in this episode, reported at length in Humboldt's journals and among the *Views of the Cordilleras* (though not in Humboldt's *Personal Narrative* of the expedition, which only reached the beginning of 1801): geology, topography, political economy, social analysis, meteorology, and anthropology. They are all linked together by an essentially physiological discourse, a view of nature focused on the reactions of the irritable living body to the fluctuating physical forces that make up nature. It is well known that Humboldt conceived of his tropical sojourn principally as an exercise in comparative physiology and anatomy, an opportunity to "investigate the influence of inanimate nature on the animate plant and animal creation."[2]

As the Quindío episode suggests, this physiological discourse served Humboldt not just to construe tropical nature but, just as important, to build a tropical self capable of experiencing that tropical nature. Humboldt's account of the American tropics is pervaded by physiology, especially a close attention to his own physiological and aesthetic responses to outside stimuli—a trait shared with many diarists and letter writers of the late eighteenth century. This physiological scaffolding was put in place in the decade before he left for America, as an experimental philosopher and Prussian official. In this chapter, I examine the function of this attention to one's own physiology through several examples. In particular, I suggest that attending to the effects of the tropics on one's own physiology (including aesthetic effects) was critical to establishing one's authority as a philosophical traveler. I will also suggest that, while the tropics exemplified the rule of nature for Humboldt, they could only perform this function in relation to other, quite different sites for fashioning self and nature: the laboratories and mines of Europe.

TROPICAL PHYSIOLOGIES

Humboldt consistently defined the tropics through physiology. For instance, when Humboldt and Bonpland made the crucial decision of which of the six thousand plants in their herbaria to present first to the Institut National of France and the public—every traveler worth his or her salt knew that few things were as effective in stimulating interest in a voyage as a few carefully chosen engravings of new and spectacular plants—they chose a palm tree, that emblem of the tropics, and one representing a new genus at that, which they named *Ceroxylon andicola,* now split into *Ceroxylon alpinum* and *C. quindiuense* (fig. 3.2). *Ceroxylon* was of extraordinary height (58 meters) and grace, but as the name they gave it suggests (*cera* = wax, *oxylon* = wood), these were not its truly distinguishing characteristics. Bonpland and Humboldt insisted on publishing *Ceroxylon* first because it offered "an extremely striking phenomenon of plant geography" and highlighted the flexibility, variability, and power of organic forces. Whereas no palms had ever been documented above 974 meters, this one first occurred at 1750 meters, "equal to the elevation of the Puy-de-Dome summit or the Mont-Cenis pass." (Note the juxtaposition of palms and ice, tropical and alpine landscape: here was a palm, Humboldt told his colleagues, found principally among the snowy summits of Tolima and the Quindío Pass.) Bonpland reported observing this new genus in abundance above 2800 meters, just 800 meters below the line of seasonal snow and in a region where mean temperature is at most 20° C, a good 17° beneath the known limits of other palms. As their name for the new genus announced, a remarkable chemical physiology went along with this remarkable geography: most notably, the palm was covered with a thick white waxy coating, evidence of a novel and potentially useful play of chemical affinities. The sample retrieved by Humboldt and Bonpland was currently undergoing analysis in the laboratories of the École Polytechnique.[3] In short, Humboldt and Bonpland chose to introduce the Institut and the public to the fruits of their voyage through a plant that was visibly iconic and beautiful but truly distinguished for its invisible physiology.

Mosquitoes were emblematically tropical in their own way. They appear constantly in Humboldt's expedition journals and yield a long dissertation in the personal narrative (the *Relation historique*) of the voyage. Indeed, in the journals, they crowd out many of the more obvious hardships of tropical river travel. For instance, Humboldt's report of his expedition

FIGURE 3.2 *Ceroxylon andicola,* from Alexander von Humboldt and Aimé Bonpland, *Essai sur la géographie des plantes, accompagné d'un tableau physique des pays équinoxiales* (Paris, 1807) (by permission of Cambridge University Library)

down the Orinoco and Rio Negro during April and May features its share of crocodiles, snakes, and jaguars leaping on canoes from overhanging trees, near-drownings, and day after day of eating nothing but manioc, banana, and occasionally ants and earth, but mosquitoes dominate.

> In Guyana, where mosquitoes darken the skies, your head and hands must always be covered, so that it's almost impossible to write in daylight. You can't hold a pen still, so fiercely does the venom of these insects sting. All our work had to be undertaken in an indian hut, where not one ray of sunlight penetrated and you were forced to crawl around on your stomach. You suffocate from the smoke, though you suffer less from the mosquitoes. In Maipures, we and our Indians stuck ourselves in the middle of the waterfall: the river is deafening, but the mist drives away the mosquitoes. In Higuerote, people bury themselves in the sand at night, so that only the head sticks up and the entire body is covered with 3–4 inches of earth. I'd have thought it a fable if I hadn't seen it for myself.[4]

Likewise on the Rio Magdalena, which Humboldt traveled for more than four months in 1801. On the Magdalena, Humboldt gathered observations he had collected on the behavior of mosquitoes from different regions: their morphology, their active hours (on the Rio Magdalena you can write, but you can't sleep; on the Orinoco, forget about writing, but at least you get some rest), their aggressiveness, and especially their relative toxicity. Like crocodiles and human beings, he points out, mosquitoes differ in ferocity, toxicity, and activity, according to their environment.

This dissertation on the relative toxicity of mosquitoes in different regions led to self-experiment. Humboldt eventually began experimenting on mosquito venom by pricking himself with thorns along his arm, rubbing the wounds with ground-up mosquitoes, and comparing the resulting wounds. This is not just an instance of self-experimentation. It is also a reflection on and exemplification of the peculiar energies, powers, and labors of the tropical naturalist:

> These factors are manifested principally among travellers and foreigners navigating rivers famous for their mosquitos. To the stimulation of travel is joined that of mosquito stings. . . . The unceasing work in which they are engaged by the productions of a nature as rich as it is beautiful, becomes a new cause of debilitation for the naturalist. The poor food, the lack of wine and other stimulants, the lack of muscular exercise, the stagnant and humid air of a river valley filled with rotting vegetation . . . upset the functions of the digestive system. The stings of mosquitos, the large quantity of venom with which one is inoculated, the inflammation of the skin, the heating of the blood, the resulting aggravation—all this disposes the nerves, their irritability dangerously elevated, to succumb to the miasmas.[5]

Thus meditations on mosquitoes as a danger to health by unbalancing irritability become an occasion to illuminate the sensitive physiology, the peculiar animal economy of the naturalist-traveler.

In this light, Humboldt's first reports home to his friends and family sound distinctly medical. Attention to and reflection on tropical physiology—the physiology of the tropics and Humboldt's own physiology in the tropics—permeates his correspondence and diaries from the moment he takes his first steps on American soil. In a July 1799 letter to his brother Wilhelm, describing his first week on the American continent, Alexander gives a vivid picture of sensory overstimulation and nervous exhaustion. Prodigies of organic development compete for the naturalists' attentions and leave them at a loss for where to begin—coconut palms fifty to sixty feet high, banana palms and a host of other trees with huge leaves and hand-sized, perfumed flowers, vibrantly colored birds and fish.

> We've been running around like madmen; in our first three days we were unable to identify anything, because we keep tossing aside one thing to grab another. Bonpland assures me that he will leave his senses, if the marvels don't stop. Still, more beautiful even than these individual miracles is the impression made by this powerful, luxurious and yet so gentle, cheerful vegetation. I sense that I will be very happy here, and that these impressions will often raise my spirits in the future. . . .
>
> NB: I am not at all worried because of the torrid zone.[6]

After describing the not unpleasant assault of tropical nature on his senses and sanity, Humboldt frequently reassured his correspondents of his physiological and mental resilience and intactness. For all the hardships he suffered, Humboldt repeatedly affirmed his fitness for the tropics, indeed, his tropical nature. "Despite moving constantly between humidity, heat and mountain frost, my health and happiness have visibly increased since leaving Spain. The tropics are my element [*die Tropenwelt ist mein Element*], and I have never been so constantly healthy as in the last two years. I work a great deal, sleep little. While making astronomical observations I am frequently subjected to the sun, hatless, for four or five hours at a stretch. I have sojourned in cities where the dreaded yellow fever raged, and never, ever had so much as a headache."[7] The key to his tropical self-construction is that he has melted into the tropical play of chemical affinities. He becomes superhuman in the tropics (as in the mines). His physiology and tropical nature are one.

TROPICAL PLANT PHYSIOGNOMY AS AESTHETIC EPIPHANY

Humboldt created the disciplines of geography and physiognomy of plants in part to enable further attempts to test and display one's tropical physiology. Humboldt famously expounded at length on the effects of landscape on *l'homme sensible,* the sensitive viewer, principally through the form, mass, and color of its plant life: the "physiognomy" of its vegetation. Humboldt interpreted plant physiognomy in terms of physiological forces. One could recognize in the physiognomy or morphology of vegetation, in the expansion and intensification of forms that were held to fixed types by "the primeval force of nature," the increase in vital force from the poles to the equator. Contemplating the vegetation clothing the banks of the Rio Magdalena during his 1801 ascent of the river, he described an aesthetic epiphany: "The splendor and fullness of the vegetation, the sheer size of the gigantic Bombax, Anacardia, and Ficus, give the banks of the Magdalena a grand, solemn, serious character. This character is entirely unique to equatorial regions, between 0 and 9 degrees. The horde of intertwined plants, the huge, succulent leaves of Heliconia and Cana which cover the ground, completely fill any empty space."[8] Nearly half a century later, Albert Berg repeated Humboldt's 1801 ascent of the Rio Magdalena and the Andes in a sort of aesthetic pilgrimage to see and experience what Humboldt did and what he had conveyed in his writings. (fig. 3.3 shows one study from the English translation of Berg's *Études physiognomiques,* the record of his 1848–50 expedition). Such studies attempted to capture the associations of plant forms that, according to

FIGURE 3.3 *Primeval Forest at an Elevation of about 7,000 Feet; in the Distance Mount Tolima,* from Albert Berg, *Physiognomy of Tropical Vegetation in South America* (London: Colnaghi, 1854) (by permission of the British Library)

Humboldt, especially determined the particular aesthetic character of a landscape. They did not merely aim for picturesque effects, however; in their ability to convey a physiognomy, they were also demonstrations of physical law and the "co-operation of forces." In a letter that Berg used as a preface, Humboldt praised Berg's depictions for their ability not only to excite the viewer's emotions through the grandeur and wild profusion of their objects, but also to raise the viewer's intellect to the concepts of physical geography, "to remind [us] of an intimate connection between the distribution of forms and climatic influences, depending on the elevation of the plateaux and the latitude of the locale." To this end, each of Berg's lithographs was accompanied by a botanical guide supplied by Ferdinand Klotzsch, director of the Royal Botanical Gardens in Berlin and Humboldt's close friend. In figure 3.3, which depicts a clearing beyond which we see the volcano Tolima, the dense vegetation is characterized by species of *Heliconia,* the waxy trunks of the Andean palm *Ceroxylon,* the aborescent fern (*Cyathea*), the young fronds of *Oreodoxa frigida,* framed by the larger trunk of a *Urostigma,* festooned with lianas and epiphytes.

The vocabulary of expansion and contraction with which Humboldt

FIGURE 3.4 *An Goethe,* frontispiece to Alexander von Humboldt, *Ideen zu einer Geographie der Pflanzen, nebst einem Naturgemälde der Tropenländer* (Tübingen: J. F. Cotta, 1807), with an engraving by Bertel Thorwaldsen (Beinecke Rare Book and Manuscript Library, Yale University)

describes the geography and physiognomy of plants derives from Johann Wolfgang von Goethe's *Metamorphose der Pflanzen* of 1790. "Wherever organic matter finds room, it expands (Goethe, Metam.), and when constantly stimulated by sunlight and humid warmth, only inner conditions (fruit and flower) set this organic expansion an end."[9] Humboldt supplied a dedication to Goethe in the German translation of the *Essay on the Geography of Plants,* ornamented by an elaborate vignette engraved by Goethe's favorite sculptor, Bertel Thorwaldsen (fig. 3.4). Humboldt read and commented on Goethe's long botanical poem in 1791, as he was attempting to make sense of the strange forms and physiologies of lichens and fungi found in the dank mines of Freiberg. In equatorial America, he gave Goethe's morphological terminology of expansion and contraction a physiological gloss. The multiplication and transformation of a basic form (the leaf organ) that Goethe had seen in development of the individual plant, Humboldt saw in the mass of plants as they developed from the poles to the equator.

As both Berg's aesthetic pilgrimage and Humboldt's debt to Goethe suggest, engaging in plant geography was also an opportunity to reflect on the depth of human feelings and to cultivate and share those feelings through plant physiognomy. There was a continuity between physiological sensitivity and aesthetic sensibility. Humboldt's manifesto for the discipline, the *Essay on the Geography of Plants* of 1805, concluded with an extended argument for a deep connection between the evolution of human feelings (in the individual and in the species) and plant geography, through the physiognomy of plants. Plant physiognomy offered, in turn, the op-

FIGURE 3.5 *Pflanzenphysiognomik der Tropen,* from Karl Friedrich von Martius, *Die Physiognomie des Pflanzenreiches in Brasilien* (Munich, 1824)

portunity to cultivate one's spirit, and this epiphany of the human spirit was best experienced in the tropics or, at least, through the image and idea of the tropics conveyed through the skills of painters. This accounted for, according to Humboldt, the "longing for the Tropics" (*Sucht nach den Tropen*) that characterized civilized Europe. Humboldt began his own expeditionary and spiritual autobiography in this longing for tropical nature—born of his 1791 encounter with the paintings of Sydney Parkinson and William Hodges in England, an episode that begins both the *Personal Narrative* and the second volume of *Cosmos,* devoted to the human spirit's hunger for nature (see also chap. 2). A steady stream of Germanic travelers returned from the tropics with exercises in the physiognomy of tropical vegetation to present to their genteel patrons, as for instance in the plant physiognomic studies that conclude Carl Friedrich von Martius's *Travels in Brazil,* published in 1824 (cf., from Martius's *Pflanzenphysiognomik der Tropen,* fig. 3.5). Martius presented these physiognomic studies of tropical vegetation to his Munich patrons much as Humboldt had to his aristocratic Berlin audience twenty years earlier, though in quite different political circumstances. The "extremity of organic development" (in terms of profusion, brilliance, texture) that characterized tropical vegetation reminded the audience of their own superiority to physiology, the victory of culture over nature only possible in temperate climes. Tropical peoples were condemned to remain mired in nature, while the peoples of Europe contemplated that nature in their glass houses.

THE EIGHTEENTH-CENTURY CULTURE OF SENSIBILITY

This construal of the tropics as an extension of the philosopher's sensibility—Humboldt's view of the world through the physiology of nervous stimulation and the concomitant attention to one's own nervous physiology—grew out of a late eighteenth-century culture of sensibility that prized attention to, and display of, feeling as a sign of natural virtue, contrasting with the artificial distinctions of society. Primarily studied as a literary genre referencing Laurence Sterne's "man of feeling," Diderot's "homme sensible," or Lessing's "menschliche Mitleid," the literature of sensibility had gained increasing attention as a social phenomenon, both describing and participating in particular historical practices of reading and display. Sentimental literature not only represented and celebrated the experience of being moved but also provided an opportunity to display and discuss emotion in late Enlightenment society.[10] Pictorial art played an important role in the culture of sensibility, and landscape vignettes became essential accompaniments to any good sentimental narrative. The works of Bernardin de Saint-Pierre came copiously illustrated with high-quality engravings, opportunities in themselves to shed a tear.

Humboldt was thoroughly schooled in observing and retailing his own inner states. He wept over Bernardin's *Paul et Virginie* and wrote about his tears to three different correspondents. He was given this book and others in the genre by his tutor, the educational reformer Joachim Heinrich Campe, who himself authored an extremely popular, sentimentalized translation of Defoe's *Robinson Crusoe* that cast the work in the model of Rousseau's *Émile*. Campe was also responsible for "anglicizing" the landscapes at Tegel, the Humboldt estate on the edge of Berlin. Humboldt's teens and twenties were a series of passionate friendships, lonely wanderings, and isolation in the middle of high society, which he reflected on at length in his youthful letters. Rousseau-like solitary promenades led Humboldt to find consolation and even friendship in the plants of the Berlin Botanical Garden. Ottmar Ette has aptly described the entire corpus of Humboldt's Jugendbriefe as an *éducation sentimentale*.[11]

The late eighteenth-century philosophical and literary discourse of sensibility drew support from contemporary medicine and psychology. The pursuit of the human mind through medicine and physiology was a common feature of Enlightenment philosophy, from the experimenters of the Royal Society and the physician John Locke through Montesquieu,

Maupertuis, Hartley, and Diderot, to early Romantics like Thomas Beddoes and Erasmus Darwin.[12] The discovery of novel powers in living matter in the eighteenth century supplied both a didactic example of the primacy of experience over theory and an important doctrine in accounts of the sensory origins of human reasoning. Albrecht von Haller's discovery of the "irritability" (*Reizbarkeit*) of living tissue, Caspar Friedrich Wolff's detection of a "nisus formativus" in the growth of a chicken embryo, and Abraham Trembley's display of the regenerative powers of the hydra all served as lessons in experimental philosophy but also as support for an epistemology based on sensation. Barometers, thermometers, electroscopes, and other instruments that claimed to reveal subtle fluids occupied a key place in this culture of sensibility. William Hogarth's prints satirizing Methodism, foppishness, and sexual license all feature barometers and thermometers indicating the vaporings and miasmas afflicting the writhing masses.[13]

LIFE AS AN ENDLESS CHEMISTRY: CHEMICAL PHYSIOLOGY

Humboldt took this entire culture of sensibility with him into his private laboratories at Tegel and Bad Steben. Humboldt's tropics were born in these cold, drafty places, as he crafted the persona of an experimental natural philosopher. In his twenties, a decade before he set off for the Americas, Humboldt began developing techniques for displaying and experimentally manipulating the "vital force" or "irritability" of organic matter—that property at the root of sensibility. He adopted the techniques of galvanic stimulation—the application of metals to muscle and nerve tissue to produce convulsions and sensations—and used them to elucidate a general chemical physiology, or "vital chemistry," that enabled the experimenter systematically to measure the effects of various forces and substances on organized matter. How would a variation of a tenth of one percent in the concentration of atmospheric oxygen affect the irritability or "vital force" of a bullfrog's thigh muscle? Would a strong magnet amplify the vital force of a nerve in the same way that it amplified the propensity of certain chemical solutions to crystallize? These questions could be answered by gauging the irritability of the organic matter in question by its galvanic response under differing conditions. In the future, such techniques promised great things for materia medica and, especially, the reformation of medicine. More immediately, however, Humboldt's experimental physiology demonstrated the dynamic complexity of physical nature and his own (the naturalist's) sensitivity to it.

Once Humboldt was able to produce and modulate artificially the contraction of a muscle fiber through the stimulation of a nerve, it became easy to see these organic tissues as themselves extremely sensitive instruments capable of detecting the slightest variation in the environment. One series of experiments published in the *Chemische Annalen* in 1795 was devoted to showing how a properly prepared muscle fiber might serve as a "living anthracometer."[14] Especially nerves. Humboldt assiduously pursued experiments demonstrating that nerves radiate *sensiblen Atmosphären* that wax and wane with the vitality of the organ. He used the phenomenon to stress the subtlety and sensitivity of the nerve's irritability and the corresponding precision with which the causal nexus that is human character must be analyzed.

The human being was only the most complex, flexible, sensitive of these concatenations of vital forces: "Were we permitted to speak of a pre-eminence of physical human nature above animals and plants, we would have to place this pre-eminence in our more subtle sensitivity [*Erregbarkeit*] or irritability, in that simultaneous receptivity to the stimulus of ideas and all stimuli of the external sensory world. Affecting everything, and affected by everything, Man is all together the middle-point of creation, and the degree of his sensitivity climbs with civilization [*Bildung*] itself."[15] The natural philosopher was not exempt from this physiology, then. On the contrary, he exemplified it. Humboldt demonstrated this physiology on himself, frequently in private and in society, and less often and at less length in print. He galvanized his own gums, eyeballs, orifices, and, occasionally, skin lesions made for the purpose with blistering plasters, in order to examine his own vital forces. Occasionally, Humboldt placed frogs' legs and other classic galvanic indicators in circuit with his own deltoid muscle.

This was part of a more widespread practice of experimenting on oneself with opium, alcohol, and nitrous oxide and recording the results. Humboldt reported these experiments dispassionately and clinically, not at all sensationally, noting which combinations of armatures and connections produce which tastes, colors, pains, and sounds, as well as muscular contractions. He assured his readers that while he was the first physiologist he knew of "to try to produce muscular movements by denuding nerve fibers on myself . . . I have been amply compensated for my self-inflicted pains by the gain in new experiences [on the effects of nervous irritation]."[16] He was displaying not himself but a general power of organized nature, which the experimental naturalist manifested to a particularly high degree. The experimenter particularly displayed this power in his ability to multiply and combine experiences, the natural power of the frog's leg raised to self-conscious scientific method through the application of precise instruments.

MINES AS SITES FOR CONSTRUCTION
OF THE TROPICS

Much of Humboldt's construction of the tropics was actually achieved by delving beneath the earth's surface, in mines. The mine is the concrete historical site of Humboldtian sensibility. In the 1790s, Humboldt had cultivated his philanthropic sensibility not just as a physiologist but as a director of mining and salt making in the Prussian principality of Ansbach-Bayreuth. As soon as he began making descents, Humboldt discovered that the interior of the earth had its own strange and dramatic meteorological and organic history, that in fact a mine was the perfect place to study the conflict of physical, chemical, and organic forces, including those of human industry. Humboldt's physiological investigations began with experiments on mine vegetation, examining the effects of subterranean gases and light deprivation on plants. These experiments on the "chemical physiology of plants" were first published as an appendix to a "subterranean florilegium," the *Flora fribergensis* or "Freiberg Flora" (1793). Like many of his German contemporaries, Humboldt used the terms "geognosy" or "geology" in their general philosophical senses to indicate that science which treated the earth historically, as the product of dynamic processes of formation. Subterranean flora, a botany embedded in the earth, made this subterranean "geography" particularly concrete and led Humboldt to describe a "subterranean meteorology," a maelstrom of forces and elements underlying the apparently static earth. Humboldt's physiological sensibility (and the measurement methods that went with it) revealed beneath the earth a landscape—a geography and physiology of plants, a meteorology, as well as a human drama—every bit as romantic and sublime as the tropical landscapes he would retrieve from America.

Thus, if experimental physiology supplied ways of displaying and cultivating the explorer's sensitivity to nature's essential dynamism, the particular political and economic locus of that sensibility was the Prussian mine. This was a social as well as a physical sensibility. To friends who urged him to conserve his fragile health, Humboldt wrote of riding fifteen hours a day, of spending whole days in the noxious, lightless atmospheres of underground galleries, and of winning the trust of his miners through his apparent indefatigability—"they think I must have four arms and eight legs."[17] In one case, assistants discovered Humboldt's lifeless body sprawled on the floor of a particularly miasmatic gallery, into which the intrepid mining official had descended to test a new miner's lamp. Before losing consciousness, Humboldt did manage to notice that his lamp still burned, defying the foul mine air when Humboldt's body could not (fig. 3.6).

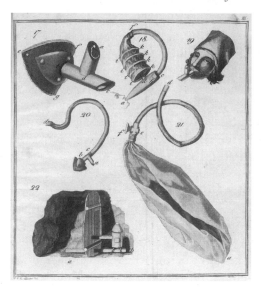

FIGURE 3.6 *Respiratory Apparatus and Lamp for Miners,* from Alexander von Humboldt, *Ueber die unterirdischen Gasarten, und die Mitteln, ihren Nachtheilen zu vermindern: Ein Beytrag zur Physik der praktischen Bergbaukunde* (Braunschweig: Schulbuchhandlung, 1799) (by permission of Cambridge University Library)

The sensibility of the scientific traveler was thus the same as the sensibility of the director of mines. Indeed, it turns out that traversing the Quindío Pass was much like descending a mine shaft, meteorologically, physiologically, and economically. This equivalence is made explicit in the description of the episode in the *Vues des Cordillères,* in which Humboldt describes the Quindío path as an "open gallery," and more completely in the *Political Essay on the Kingdom of New Spain,* in which he criticized the uneconomical expenditure of vital forces in the Mexican mines.[18]

This was no mere sentimental display of working-class solidarity or self-martyrdom. Instead, the drama and landscape of the mines gave Humboldt an opportunity to exercise and demonstrate a statesmanlike sensibility. Humboldt's understanding demonstrated the superiority of mind over matter, spirit over mechanical constraint. Humboldt's miners may have been terrified by the monstrous vegetation and deadly exhalations of the mines, but (even when physically overpowered by these very gases) Humboldt understood this subterranean physics as part of a historical struggle of forces that ultimately lead to civilization. The lamp burning in the noxious gallery next to Humboldt's unconscious body showed that.[19]

This precise sensitivity to "the reciprocal action of forces," mediated by barometers, eudiometers, thermometers, and anthracometers, enabled Humboldt to ameliorate conditions in the mines and spare the vital forces of Prussia's miners. This was true in two senses. In a direct sense, Humboldt's grasp of the interaction of force permitted him to manage and control the forces at work in the mines: he developed a miner's lamp and respirator, liberating work from the constraints and dangers of working in a dark miasmatic environment, enabling work shifts to be lengthened and standardized. It was also true in a deeper sense: to appreciate the earth as the product of a complex interaction of forces, to be sensitive to the historical, dynamic character of nature, was also to possess a humane sympathy for the suffering of human labor and a profound awareness of civilization as the product of struggle against nature. Thus, in a 1797 letter to the *Annalen der Physik* announcing that he had detected a strong magnetism in a local rock stratum (perhaps the remnant of some ancient shift in the orientation of the earth's magnetic axes), Humboldt also announced the opening of a fund for poor miners, using proceeds from the sale of samples of the magnetic deposit.[20] To grasp the dynamic character of nature was humane in both senses, and it was an epiphany rooted in the German mines, though subsequently exemplified in the American tropics.

If Humboldt helped establish the tropics as a privileged site for specific kinds of European experiences, he did so as much by representing himself as by representing his surroundings. His aestheticization of the tropics served to make the tropical landscape the paradigmatic site where Europeans might realize and display their "European-ness." I have tried to show here that this privileging depended on a common late-Enlightenment culture of observing and reflecting on one's own sensibility and that it was modeled on heroic elaborations of his own sensibilities in the laboratory and in the mines of Europe. The relationship between feeling and measurement, between aesthetic appreciation and exact science in Humboldt's work, is therefore not accurately termed a synthesis. Instead, feeling and measuring were closely identified, even continuous activities, and the display of "tropical" sensibility was an integral part of Humboldt's credibility and authority as a scientific observer.

"The Struggle for Luxuriance": *William Burchell Collects Tropical Nature*

LUCIANA MARTINS AND FELIX DRIVER

As to the Botanical riches of this country (at least, what little I have hitherto seen) you cannot form an adequate idea, even though you pictured to yourself all the fine plants of our hot houses growing wild and clothing the hills and vallies with their utmost luxuriance. . . . What pleasure should I not feel, in conducting you through forests where lofty trees decorated with garlands of Begonias, Passifloras, and trunks and branches covered with ferns, Pothos, Tillandsias, Bromelias and a multitude unknown to me, are intermingled with graceful Palms; where birds of new forms and glittering metallic colours, and Butterflies of the gayest hues, seem to people the strange scene: where mountains and rocks, forests and rivulets, and plants of remarkable forms, combined in the most picturesque manner, tempt one to turn often from Natural History to Painting.

William Burchell to Richard Salisbury, 11 August 1826

In March 1825, William Burchell set sail from Plymouth, accompanying a diplomatic mission to Brazil. By the time he composed the picturesque view of tropical nature which forms the epigraph to this chapter, Burchell was already an experienced naturalist. In addition to his travels in Brazil (which were eventually to occupy five years), he had spent five years botanizing on the island of St. Helena (1805–10), during which time he was elected as fellow of the Linnean Society, and another five in Southern Africa (1810–15). During his travels, Burchell used his considerable drawing skills, together with his scientific expertise, to record the features of numerous landscapes, peoples, flora, and fauna. He also collected a large number of botanical, zoological, and geological specimens: indeed, the unpacking, relabeling, and repacking of his collections was to occupy most of the remaining thirty-three years of his life following his return from Brazil. Tropical nature was "almost too luxuriant," he confided in William Hooker in 1826.[1] Soon after his return to Britain in March 1830, he reiterated the theme.

You, who are so great an example of industry, complain also of over-whelming collections, & feel the necessity of manual help. But I have nowhere beheld an herbarium so large as my own; and, added to this, I can-not bring my mind to abandon any branch of natural history for the sake of giving more time & attention to any one in particular; although I know this is wrong and can never lead to perfection in any. Still the contempla-tion of the whole system of created objects is so fascinating that is very difficult to turn away from all but a few.[2]

Burchell's activities as a collector, and in particular his sense of the over-whelming scale of his tropical collections, do not always, or easily, con-form to now-standard arguments about the relationship between natural science and the ordering of the world. David Mackay, for example, has ar-gued that the "first task of the Banksian collectors was therefore to tran-scend the chaos of the lands they visited and reduce the natural world of empire to order."[3] This, at least, was the view from Soho Square and the Royal Botanic Gardens. But not all collectors were in the service of Joseph Banks, and they did not all have identical projects in view. In their jour-neys into the tropics, sometimes self-supported, sometimes on behalf of others, commercial, colonial, and scientific interests were mixed together with theological, moral, and aesthetic motivations. In the case of William Burchell, no ordinary plant collector, the inability to systematize a knowl-edge of tropical nature may also reflect wider challenges to the frame-works of natural science itself: the sheer novelty and variety of the tropics could not always be easily accommodated.

By the time of Burchell's travels, the image of the tropics as a zone of circulation, a portion of the globe within which plants could be ex-changed from one place to another, was already well established. The global transplantation of species between different tropical territories in the Americas, Africa, Asia, and the Pacific had begun in earnest with the Portuguese in the fifteenth century.[4] In the process, existing networks of pharmacological trade in the East were linked to West Africa, the Carib-bean, and Brazil. Subsequently, botanic gardens were established in Re-naissance Italy, following a revival of mystical and medical interests in the study of nature, accompanied by the introduction of exotic plants in Eu-ropean lands.[5] During the seventeenth century, the Dutch set up a botanic garden at the Cape of Good Hope, primarily to grow vegetables for the ships of the Dutch East India Company. The growing interest in botany was marked by the preparation of the *Hortus indicus malabaricus* (1678–1703), a major project initiated by the Dutch colonial administration in India, Hendrik van Rheede tot Drakenstein. The *Hortus* depicted a wide range of cultivated and wild plants: grasses, ferns, and mosses, as well as coconuts, bananas, and papayas.[6] As Grove suggests, this work is better seen as a compilation of Middle Eastern and South Asian ethnobotany than a uniquely European affair.[7] Similarly, the Swedish East India Company's overseas trading operations provided Linnaeus with a worldwide network through which to organize a program of collecting across the globe. This knowledge came at a price: it has been estimated that "a third of the Linnaean disciples died during the course of their expeditions."[8]

In addition to their medical and scientific uses, tropical plants came to be grown for their aesthetic value. In Britain, the cultivation of exotic species became an important part of the popular culture of gardening.[9] During the eighteenth century, artistic depictions of botanical specimens were much in demand within polite society, including Mark Catesby's *Natural History of Carolina, Florida and the Bahama Islands,* published in 1747.[10] More directly political associations were evident in a later work, James Thornton's finely illustrated *Temple of Flora,* the plates for which were published between 1797 and 1807.[11] In a plate titled *Cupid Inspiring the Plants with Love,* Cupid aims her arrow at the *Strelizia reginae,* the bird-of-paradise, a specimen transported to Kew in 1773 and named in honor of Queen Charlotte by Joseph Banks. The vision of tropical nature presented here is idealized and exoticized, a setting for a celebration of fertility and power. Thornton's patriotic rendering of a specifically British vision of tropical nature can readily be seen as the aesthetic counterpart to the Banksian worldview.

Nicholas Thomas has questioned the emphasis placed on imperial designs in recent writings on the history of exploration and encounter. "It has become increasingly evident," he argues, "that the present range of approaches exaggerates and reinscribes precisely those western hegemonies they wishfully challenge. . . . The tendency is to insist upon the will to dominate in imperial culture, science, and vision, without investigating the ways in which the apparatuses of colonialism and modernity have been compromised locally."[12] The argument here is not that empire is unimportant but that its effects were not always predictable. In this spirit, we are concerned in this chapter less with imperial designs and more with the practices of naturalists and collectors in the field, in particular sites. We focus here particularly on the ways in which images of tropical nature may reflect, or translate, the experience of collecting, its disappointments as well as its successes. The conduct of fieldwork is of course a vital part of the practice of botany, as much as surveying or the art of travel, in which observations, notes, sketches, and collections made "on the spot" provide the raw material for further study (cf. chap. 2). Fieldwork entails encounters with the unknown; it is not simply an act of appropriation.[13] While travelers frequently interpreted what they saw in the light of their own preconceptions, there are moments when the experience of novelty breaks through the surface of what is expected.[14] In the process of encounter, as Gillian Beer puts it, "knowledge-as-experience and experience-as-knowledge are transformed as they move across, beyond and back and forth from field to field to field."[15]

Naturalists, and other travelers, attempted to comprehend the experience of tropical difference in various ways. They were instructed not only to survey but also to listen, smell, touch, and sometimes taste the unknown. But the fruits of knowledge were not always sweet. In order to situate our account of William Burchell's travels in Brazil and his graphic depictions of the forms of tropical nature, we shall first consider the experiences of three relatively lowly plant collectors in the tropics. While ostensibly a more independent traveler, with access to a variety of resources to sustain him in the course of his journeys as well as on his return, Burchell too was far from in control of his own collections.

PLANT HUNTERS AND BOTANICAL TRAVELERS

William Burchell was by no means alone in traveling to Brazil in search of specimens of natural history. In the wake of the transfer of the Portuguese court to Rio de Janeiro in 1808 and the ending of the Napoleonic Wars in 1815, a succession of European scientific missions made their way to Latin America, hoping to capitalize on new opportunities to collect tropical nature.[16] As Joseph Banks made clear in a letter to the superintendent of the Royal Gardens at Kew,

> the arrival of the definitive treaty with France and the certainty that before any collection can be ready to be sent home Ships will sail as they were used to do without being subjected to any uncertain delays makes me anxious to see the establishment of foreign collectors resumed and the more so as the Emperor of Germany who has formerly freighted Ships at an immense expense and sent well educated Botanists to collect for his Garden at Schoenbrunn (the only rival to Kew that I have any fear about) will no doubt resume the business of improving it.[17]

In a spirit of imperial rivalry, Banks sent out a plant-hunting expedition to Brazil, consisting of Allan Cunningham and James Bowie, two young employees at Kew. Their lowly status was clear from Banks's instructions: he insisted that they were not to "take on themselves the character of 'gentlemen'" but, rather, "to establish themselves in point of board and lodgings as servants ought to."[18] At Rio de Janeiro, they were to collect seeds and specimens of plants not represented at Kew, including species of the genus *Fuchsia* native to Brazil, while taking care to avoid the perils of a tropical climate: "You are by no means to presume on the vigor of your youth and the [strength] of your constitution," wrote Banks, "but diligently to conform yourselves to the practice of the Natives in avoiding the dangers of the Climate and the hazard of visiting unhealthy districts." The collectors were instructed to keep a daily journal of all their proceedings, "especially noting the Soil situations and exposure of every remarkable Plant," and were "not in any Case to allow any Person whatever to receive any part of the Seeds or any Plant or Bulb, collected by you while you continue in your present Employ."[19]

The collectors' journal, which survives at Kew, faithfully records—predictably—that they followed these instructions diligently.[20] In February 1815, the pair wrote dutifully to Banks of their desire to collect as many tropical plants as they could, "as much that Britain would stand unrivalled in Botanical productions for ages."[21] For these tropical collectors, as for many of their Northern European contemporaries, Brazil was a "virgin country" to be explored, notwithstanding its colonization by the Portuguese for three hundred years.[22] But there were numerous obstacles to overcome, from the unfamiliarity of the various forms of nature they encountered to the unruly life of the city itself. Eventually, however, their botanizing in and around Rio itself transformed their sense of the landscape to something less threatening. On leaving Rio, Cunningham wrote, regretfully:

> That both of us should have left Brazil (at a point when we had acquired a knowledge of the language, of the country; had become accustomed to the modes of traveling and habits of the people; had accumulated friends, Portuguese and some English, all of whom found infinite pleasure in aiding us in our pursuits whereby we might have formed very extensive collections of the brilliant stores of that virgin country at much less expense than we did when perfect strangers) is a matter, in truth, I cannot too much regret.[23]

While Cunningham liked to think of himself as a "botanical traveler," from the perspective of Banks he was merely a functionary of Kew. The same might be said of the plant hunters dispatched by the Horticultural Society of London (founded in 1805), which pursued a smaller-scale program of collection and cultivation.[24] Like the Kew collectors, these men received detailed instructions for their task, and they were also dependent on the support of the admiralty. One such collector was John Forbes, who accompanied an 1822 surveying expedition to East Africa on HMS *Leven* commanded by Captain Owen, one of the leading scientific figures in the navy. His detailed instructions from the Horticultural Society of London stipulated that he was under the direction of the captain during the entire voyage. There followed a list of plants and seeds to collect, arranged in different categories: horticultural, ornamental, and medicinal. Curious specimens of fruits were to be bottled in spirits; capsules, and seeds of plants preserved for exhibition in the society's apartments. Forbes was asked to procure, in particular, seeds or plants of fruits native to hot climates and to record "every circumstance relative to their habitats in a wild state and to their cultivation or necessary treatment, where in gardens."[25] A set of more detailed instructions included requests for collections of specific species from the ports where the vessel would call: chestnuts from Portugal, citrus fruits from Madeira, Teneriffe, and Cape Verde Islands, bulbs from Rio de Janeiro and the Cape.[26] Forbes was furnished with many items of equipment, including boxes, trunks, pencils, paper folders, millboards and brown paper for his herbarium, magnifying glasses, compasses, knives, scissors, bottles for fruits, spirits for preserving fruits and insects, dissecting instruments, bags of shot, and nets. He was also provided with a collection of garden seeds, for exchange when necessary, along with letters of introduction, a small library, and technical instructions on other aspects of natural history.[27] Finally, like Cunningham and Bowie, Forbes was required to keep regular journals of his proceedings and observations.[28]

Forbes's letters and journals trace his experiences on the voyage, his impressions of tropical scenery, and his activities as a would-be botanical traveler.[29] On 1 May 1822, the *Leven* entered the harbor of Rio, which Forbes described in terms that echoed the response of other visitors in this period: "Magnificent," "a grand and noble landscape," "rich and beautiful scenery," "one of Nature's Botanic Gardens."[30] At Rio, he reported an encounter with the Russian consul Georg Heinrich Langsdorff, who was at that time organizing a major scientific expedition to the interior.[31] Having spent a month in Rio, the botanist expressed regret at his lack of opportunity to travel into the interior: "This is such a luxurious climate and the objects so nume[rous] that I think if I was to remain here for seven years I sh[ould find] something new in every day's excursion, it is an inexhaustive field for the naturalist."[32] On arrival at the Cape a month later, this disappointment was only enhanced by his view of the barren landscape of Table Bay: "There is nothing particularly beautiful or interesting in the scenery hereabouts."[33] Although Forbes repeatedly requested a copy of the first volume of Burchell's newly published *Travels in the Interior of South Africa,* his employers at the Horticultural Society dismissed the volume as a "poor performance [containing] nothing that will interest you as far as any Spot you are likely to visit" is concerned.[34] Clearly Forbes had ideas, and demands, above his station.

During his journeys up the eastern coast of Africa, Forbes was engaged in "very active botanizing."[35] In June 1823, while in Algoa Bay, he reported a proposed trip to survey the lower reaches of the Zambezi River:

"I hope to return laden with the vegetable treasures of country as little known as any in the world."[36] But the harvest of this last expedition was bitter. In his excitement at the prospect of tropical riches, Forbes's powers of horticultural discernment deserted him, and he fell victim to an "ill-judged delicacy." In the words of Captain Owen, writing to his employers: "It grieves me exceedingly to have to inform you that poor Forbes is no more. He died on his voyage up the Zambezi to Senna just when he was entering on a most interesting part of his researches. It is also lamentable that his death appears (as far as it can be accounted for physically) to have happened in consequence of retention from an ill-judged delicacy whilst voyaging in a canoe."[37]

DRAWING THE TROPICS: WILLIAM BURCHELL

The field experiences of Forbes, Cunningham, and Bowie provide glimpses of the trajectories of a class of plant collectors working in the service of metropolitan science, disorientated by the initial encounters with tropical nature, but still trying to fashion opportunities for themselves (cf. chap. 6). Together, these stories help to situate the work of the naturalist who is the main focus of this chapter—William Burchell, who arrived in Brazil in 1825. There are some similarities and many differences in their experiences of collecting, and in what follows we focus especially on Burchell's cultivation of a new sensibility, that of the philosophical naturalist, specifically through the art of drawing.[38]

Although instruction manuals for botanical and zoological collectors stressed the importance of recording perishable data in words and sketches made "on the spot," this was an activity very often neglected. In her study of early nineteenth-century British zoology, Anne Larsen distinguishes between two categories of such data: information relating to the content of a living being, such as color, shape, and habits, and information on its context, that is, where it was located, describing the characteristics of its natural habitat.[39] As she contends, the "detailed examination of landscapes was one of the most fundamental aspects of natural history; instruction manuals usually taught their readers how to see, gather and record objects that conveyed a sense of an area's particular natural features and characteristics."[40] To put instructions into practice, however, was not at all simple. Apart from an abundance of free time—for collecting and drawing are both time-consuming activities—sketching also demanded trained hands and eyes. It is worth noting here that most of the illustrations of exotic flora in contemporary botanical works were drawn from species cultivated in Europe.[41] The art of drawing from nature, especially in the context of tropical travel, was a skill developing alongside the emergence of new paradigms in natural history.[42]

In the new programs of natural history associated with Humboldtian science, for example, learning to observe the natural world was a complex activity, which required more than just collecting: for the truly philosophical naturalist, theoretical reflection gave the practice of collecting its real meaning.[43] Humboldt himself traveled in Latin America from 1799 to 1804, in the company of botanist Aimé Bonpland, and a thirty-volume account of their expedition was published in France between 1814 and 1825 (the first English translations appeared between 1814 and 1829).[44] Humboldt was deeply conscious of the mediated nature of perception: "Lucky are those who travel the world to see it with their own eyes, trying to understand it, and recollecting the sweet emotions that nature inspires!"[45] His penchant for experimentation with graphic representations in the form of thematic maps, isolines, and graphs, among other devices, suggests that he was constantly "looking for a language at once highly descriptive but also analytical."[46] At the same time, his use of images and words was conceived as a way of conveying emotions, of evoking the sensibility of the cultivated European mind seeking to comprehend the pattern of nature (see chap. 3).[47]

For Burchell, as for Humboldt, the art of visual representation—especially the depiction of landscape—was a vital tool of scientific description. At the age of fifteen, Burchell had received instruction in landscape drawing from Merigot, a French artist settled in England. His knowledge of perspective had been acquired from John Claude Nattes, a master in topographical drawing and watercolor painting.[48] As Charlotte Klonk points out, Nattes was well attuned to contemporary appreciations of the sublime in nature.[49] In Burchell's St. Helena sketchbook (now in the Kew archives), we see his early attempts to deploy his artistic skills in depicting the unfamiliar forms of tropical nature.[50] In among numerous profiles of the island drawn from the sea, on board the East India Company ship *Northumberland,* there are several landscapes and detailed drawings of natural history specimens. Two of these drawings deserve particular attention in the present context. The first depicts a hermit crab (plate 4), its delicate coloring bearing witness to Burchell's search for visual precision in the depiction of nature. In its attention to the finest of details, this drawing might be said to anticipate John Ruskin's concerns with the ideal of scientific observation in art. The second drawing is a small but remarkable sketch titled a *Group of Plantains from Nature,* dated 20 February 1807 (plate 5). The sketch includes drops of the plantain's own juice that have spilled onto the page, by accident or design. These drops are themselves

FIGURE 4.1 William Burchell, *Inside of My African Waggon,* 1820 (Oxford University Museum of Natural History)

used as a sort of evidence by proxy—"not blood but drops of Plantain juice," writes Burchell on his sketch.[51] In this way, the visual image becomes something more than mere representation: stained red by the actual specimen, the very scrap of paper itself acquires scientific value. No longer just an "illustration," Burchell's sketch provides confirmation of the authentic presence of the observer in the field, thereby affirming his credibility as a faithful witness.

In 1810, Burchell left St. Helena to take up the post of "Botanist to the Cape Colony." His subsequent four-year journey of 4,500 miles across southern Africa resulted in the publication of his best-known work, *Travels in the Interior of Southern Africa* (1822–24). Richly illustrated with engravings based on his own sketches, these volumes are peppered with reflections on the art of drawing and its usefulness for the traveling naturalist. For those visiting little-known regions, he argued, its value lay "not merely as the means of giving their friends an idea of those scenes and objects which they had beheld, but for their own gratification, and for the pleasure of a renewal of past impressions far more lively than any pen can render a written journal."[52] Nowhere was this pleasure more discernible than in the small but highly detailed watercolor that Burchell produced on his return from his travels, depicting the interior of his purpose-built collecting wagon (fig. 4.1). As he states in his notes, it took him four days of sketching and twenty-seven of coloring it, or "about 120 hours' work on the whole" to complete the work.[53] The wagon was represented by Burchell as his most precious piece of equipment: not just a mobile laboratory, in which he could read, write, draw, weigh, measure, dissect, and skin, but an instrument itself, the rotations of its wheels providing a means of calculating the distances traveled.[54] In the detailed rendering of the contents of the wagon—complete with specimens, maps, charts, and a host of instruments of all kinds—there is also evidence of a certain nostalgia for his traveling home. Back in Fulham, amid his travel journals, notebooks, and innumerable specimens, perhaps he was longing for the sense of order and control the wagon had secured while he traversed the landscapes of South Africa.

From this specially designed mobile laboratory, Burchell was able to enjoy the novelty of his surroundings, which though not tropical were certainly exotic. For him, the extensive plains of South Africa possessed "a species of beauty with which, possibly, they [European artists] may not yet be sufficiently acquainted: some beauties of its own, which depended more on the effects of aërial tints and the colouring of a warm arid country, than on richness of subject or a romantic outline."[55] In accordance with an influential strand of contemporary opinion in Britain, Burchell suggested that the traveling naturalist should adopt a more disciplined approach than the artist who regarded nature "as the medium through which he may display his art, and afford amusement" and, instead, consider his "art as the means of exhibiting nature, and of conveying information."[56]

Equipped with such a sensibility, borne out of his very specific trajectory as a naturalist and a collector, Burchell set out on his final major voyage, this time returning to the tropics to accompany Sir Charles Stuart's diplomatic mission to Brazil in 1825. Burchell's interest in Brazil may have been kindled by William Swainson's descriptions of its luxuriant scenery.[57] And in laying still more ambitious plans to visit "the interesting remains and antiquities of the empire of the Incas," Burchell may have had in mind Humboldt's detailed drawings of Aztec art. On 15 March 1825, Stuart's mission sailed from Portsmouth to Rio de Janeiro, accompanied by a retinue of officials and an artist, the brother of the painter Edwin Landseer.[58] Burchell remained at Rio for well over a year, during which time he visited Minas Gerais and the Órgão mountains. There he contemplated a much more ambitious journey, which would take him across the entire continent:

> My plan and wish is . . . to explore the provinces of St. Paulo; Goyáz; Cuyabá; and Mátto-Grosso. Then to pass into Peru, with the principal object in view, of making a short residence in the city of Cuzco, to examine

the interesting remains & antiquities of the empire of the Incas, both there and in the country surrounding the great Lake Titicáca. I know not whether I may hope to accomplish a further march on to Lima, or whether my family and particularly my dear Father would not feel uneasy at so long an absence as this would require. Having completed my observations at Cuzco, I would turn my course southward and visit the more remarkable & interesting places thence onwards to Buenos-Ayres, where my travels would end.[59]

An experienced traveler, Burchell knew well that it was "so much easier to mark out an interesting line of travels on a map than to trace it in the country itself."[60] Before venturing into the interior, he attempted to equip himself with a more detailed knowledge of the language and customs of the country, as well as devoting much time to collecting botanical, entomological, and geological specimens, making observations on astronomy, landscape, natural philosophy, and drawing. He also found some time to socialize with the European residents of Rio, notably the naturalist and traveler Maria Graham. Graham, like Burchell, was a regular correspondent with William Hooker and shared an interest in drawing from nature.[61] Burchell seems to have enjoyed her company, in contrast to other members of the British community in Rio, such as Admiral Graham Hamond (commander of HMS *Wellesley*), who had no time for her independent scientific and literary pursuits.[62] Graham's sensibility as an artist might be compared with Burchell's, as we can see in her representation of a Heliconia (fig. 4.2). Here the imposing, vividly colored figure in the foreground contrasts with the pale view of the lagoon and the Corcovado Mountain in the background. The point of view is low; it seems that Graham was kneeling when she sketched the image, as if in reverence.

Departing from Rio in September 1826, Burchell proceeded by sea to Santos, where he remained three months, and then traveled up to São Paulo, arriving there in January 1827. He spent seven months on the outskirts of the city, exploring the surroundings and making preparations for his journey into the interior. In August 1827 he set out on his travels to Goiás, where he passed the rainy season gathering together a large collection of botanical specimens. On receiving news that his father was seriously ill, however, Burchell finally abandoned his plan of going on to Peru and decided to travel directly to Pará (now Belém do Pará) on the north Brazilian coast. He sailed down the Tocantins River, reaching Porto Real (now Porto Nacional) in November 1828, finally arriving at Pará in June 1829. After a further delay, he sailed to England, arriving at Dover on 24 March 1830.[63] Having had several months to reflect on his travels in Brazil, Burchell described his experiences of tropical nature in a letter to William Hooker in a manner that managed to synthesize the sublime

FIGURE 4.2 Maria Graham, *Helico-nia*, 23 October 1824, Rio de Janeiro Portfolio (Royal Botanic Gardens, Kew Archives)

rhetoric of tropicality with the comparative sensibilities of the Hum-boldtian naturalist:

> You have from all quarters heard the most animated descriptions of luxuri-ance & richness of the vegetation of Brazil; and with them I most warmly agree. But this is become almost a fashion, & in Europe it seems the gen-eral opinion that the *whole* of that country is covered with the most magnificent forests, of gigantic growth. This idea though correct with re-spect to all maritime districts, the courses of the rivers and the greater part of the country lying under the equinoctial Line, is however not at all ap-plicable to vast tracts in the provinces of S Paulo & Goyaz. There I have tra-versed boundless plains or open regions some of them covered with fine pasture. . . . These arid groves have sometimes reminded me of the *Acacia* groves so predominant over the plains in the interior of Southern Africa— Yet it is rarely that one can compare African with Brazilian botany; their character in many particulars differ so widely. . . . When however we de-scend towards the low latitudes of Brazil the glorious magnificence of the forests is truly astonishing, and none but those who are born in the midst of them can view such imposing productions of Nature without a feeling of awe or respect. She overloads herself, and one object oppresses and smothers another in the general struggle for luxuriance.[64]

Burchell produced at least 260 drawings during his Brazilian journey, now held in the Museum Africa in Johannesburg, some of which he used as a basis for oil paintings composed after his return to England.[65] Included in these drawings is a detailed panorama of the city of Rio de Janeiro, which took him five or six weeks to complete.[66] In a letter to his mother, Burchell complained that he had to pack it up, in order to avoid the interruption caused by the number of people that came to see it.[67] Traveling naturalists like Burchell were effectively providing the people of Rio with a way of seeing their own city, which was to have a significant impact locally as well as within Europe itself. Burchell's impatience with the locals' curiosity, added to the absence of people in his panorama and indeed the panoramic point of view itself, highlights the distance he sought to establish between himself and the Brazilians, whom he characterized in one letter home as "a most ignorant & unsociable people."[68]

Burchell's representation of the Brazilian population, far from exceptional among European visitors in this period (cf. with David Arnold's discussion of Hooker's description of the Singhalese in chap. 8), was intended to highlight his own cultivated and philosophical tastes as a naturalist. His claims as a collector of tropical nature were considerable; as he wrote to William Hooker in 1830:

> The *whole* of my collections are now in England, and when I get them safe out of the custom house, I may safely assert that in this house I have an herbarium containing, at least 140,000 specimens, and perhaps much more. My American collection has more than 7000 distinct species. In every other part of Natural History I have preserved a far greater number of objects in the new Continent than during my travels in Africa; except in the class of Mammalia alone. In insects particularly I have many times more. My astronomical and philosophical observations are also very numerous; as I devoted a great portion of my time to them. My drawings are also numerous and large. All these things the fruits of singled-handed labours, I hope to have the gratification of showing you at some future day, when I shall have them unpacked.[69]

What Burchell did not acknowledge in correspondence with his fellow naturalist, however, were the labors of others that had gone into the building of the collection: in particular, the "very useful" black boy, whose loyalty was gained—so Burchell wrote to his father—under the threat of him being sold to "some sugar plantation, where he will have to work hard and remain [a] slave for the rest of his life."[70] Burchell was far from exceptional in boasting of his "single-handed labours" in building his collection. As Jane Camerini argues in her work on Wallace's travels in the Malay Archipelago later in the nineteenth century, "the contributions of local people became invisible through the conventions in particular genres of scientific writing."[71]

Burchell's collections were certainly vast, even by contemporary standards. In his manuscript "Memoranda Botanica," Burchell made a precise record of the time he spent unpacking the botanical collection, an activity that he began nearly seventeen years after his return to Britain, on 3 February 1847, and ended three years later, on 15 February 1850, noting the state in which the specimens were found at the time and an account of the quantity of paper used in rearranging them. This attention to minutiae in recording every single piece of information is also reflected in Burchell's "Catalogus Geographicus Plantarum Brasiliae Tropicae"; descriptive notes made from the living plants including a number indicating the order in which the species were collected, the geographical site, remarkable characteristics for the previously "undescribed" species, indigenous names (when available), and numbers that referred to the colors of the plants observed from nature, together with a sample sheet against which the hue of individual specimens could be calibrated (plate 6).

Burchell's detailed notebooks provide material evidence of his own labors. Had he allowed himself to purchase species from local traders, instead of collecting them himself, Burchell acknowledged that he might have assembled a still larger collection; "but this is not the way to study nature"—he insists—"it is not in cabinets that we can discover those peculiarities of genera or species, which lead to a knowledge of the beautiful harmonies of nature. I have always wished to study the works of the creation (subliminal, at least) *in situ*."[72]

If observation on the spot was the trademark of the field naturalist, the translation of what was witnessed in the field into data to be manipulated in his cabinet required the composition of reliable records in word and image.[73] Burchell had already expressed his philosophy of natural history in a passage in his *Travels in the Interior of South Africa:*

> It must not be supposed that these charms [the pleasures of Nature] are produced by the mere discovery of new objects; it is the harmony with which they have been adapted by the Creator to each other, and to the situations in which they are found, which delights the observer in countries where Art has not yet introduced her discords. To him who is satisfied with amassing collections of curious objects, simply for the pleasure of possessing them, such objects can afford, at best, but a childish gratification, faint and fleeting; while he who extends his views beyond the narrow field of nomenclature beholds a boundless expanse, the exploring of which is worthy of the philosopher and of the best talents of a reasonable being.[74]

Nevertheless, tropical exuberance did not always, or easily, fit comfortably within the theoretical "system of nature," despite the naturalist's efforts to contain it. Burchell continuously resented the lack of physical space to manipulate his collections, as well as the time he spent dealing with his own family's affairs.[75] Even in his carefully composed "Memoranda" there are traces of uncertainties: "In this and in the preceding three volumes I think that in general I have *under*-rated the heights of those trees which are taller than 10 or 12 feet; and that the heights therein noted as 15 feet must be read as 18 feet—and all others augmented in the following proportions: For 15 read 18; 20–25; 30–40; 40–55; 50–70; etcetera."

Burchell's awareness that his work could end in failure was exacerbated by his grievances against the government for what he regarded as undeserved neglect. His sense of failure, and the feeling that others were continually gaining credit for work that he had done but not published, may well have led to his suicide on 23 March 1863.[76] It seems that Burchell, too, was caught by that "commonplace bifurcation between the ecstatic profusion of tropical nature and its pervasive menace" that Hugh Raffles has identified in the collecting career of Henry Walter Bates.[77]

CONCLUSION

The boundaries between the concerns of tropical travelers encountered in this chapter—plant collectors and philosophical naturalists—were often blurred in the early nineteenth century. Natural history, as David Allen has argued, is not and never has been a purely intellectual pursuit: "Even in its most primitive manifestation, collecting, there can be a delight in shapes and colors and patterns that co-exists with the mere pleasure of acquisition or the sheer satisfaction of having the evidence of some additional knowledge."[78] In fact, very often the business of collecting demanded far more than "primitive" skills—and it certainly could absorb more time than just observing and recording. The making of a visual record of nature's forms, however, also required a specific way of seeing; and as Anne Secord remarks, drawing "functioned as a learning process only for those already aware of what they should be looking for."[79] Burchell's extensive visual archive of natural history contrasts with the absence of surviving drawings by Cunningham, Bowie, or Forbes and bears witness to a fundamental difference that was being constructed between humble plant hunters and "true" naturalists: only the trained eyes of the scientific observer knew how to observe philosophically.

The sheer range of artifacts that naturalists sent home, ranging from stuffed birds to fossils, had a double destiny: they were both commodities for private display and subjects for scientific inquiry.[80] In the private sphere, the taste for bottling up natural objects was already well established.[81] In the public realm, meanwhile, the "greatest palm house in the world" was about to be built at Kew. The connections between the search for exotic flora and the technology of modern voyaging were evident in the design of this very building: wrought iron "deck beam" was used for its main arches, conferring on it the look of an "upturned hull of a graceful liner."[82] Furthermore, as Larsen notes, there was a mutual relationship between aesthetic interests and popular science, insofar as exotic forms nurtured, and were nurtured by, shifts in fashion among the consumers of illustrated books and other commodities.[83]

When first setting eyes on tropical nature, philosophical naturalists such as Burchell and Humboldt searched for associations—in literature, poetry, and the visual arts. What Darwin called (in 1832) the "total dissimilarity of a Tropical view" was apprehended through a well-established discourse.[84] Rather than following Burchell's concern with the limits and classification of species, however, Darwin sought evidence of dynamic relations and transformation. Nonetheless, he, too, experienced something of the sense of dislocation in voyaging through the tropics, in the process of collecting, observing, drawing, and making sense of what was experienced on the spot. Even in the published, official accounts of such voyages, writes Gillian Beer, "fascination with the unfamiliar, fear and loathing, the longing for stable systems of communication, sickness, religious fervour, and the physical pleasures of exploration all pressed across and became part of enquiry and leave their traces in writing."[85] Travelers are not always in control; collections may sometimes overwhelm their makers.[86]

MAPPINGS

★ 5 ★

Dominica and Tahiti:
Tropical Islands Compared

PETER HULME

That very special reek / tristes, tristes tropiques
Derek Walcott

My aim in this chapter is to compare the ways in which the peoples and landscapes of Dominica and Tahiti have been described by outsiders. As tropical islands, Dominica and Tahiti have some general similarities: they are roughly the same size, both are mountainous, they have roughly the same population, Dominica is about the same distance north of the equator as Tahiti is south. However, my starting point is that Dominica and Tahiti, and their histories, are so obviously different in so many respects that the challenge is to find more meaningful ways of bringing them together, ways that might illuminate the nature of "tropical visions." In trying to bring the islands together, I pay particular attention to the ways in which they have been brought together over the past two and half centuries, the ways in which frames of reference have been created in which Tahiti and Dominica both have a particular place, and often a special place, the principal frame being that of the imaginative construction we have come to think of as tropicality.[1]

I

There are, of course, many differences between the two islands and their histories. Dominica first came to European attention in 1493 when Columbus sailed past it and gave it that name—its inhabitants had called the island Waitukubuli; it was nearly 300 years later before Europe became aware of Tahiti, which—after a struggle—kept its indigenous name. As far as Europeans were first concerned, Dominica was home to a savage tribe of man-eating Indians, the Caribs, while Tahiti was populated by hospitable natives who were equally free with their food and their sexual favors. As the Caribbean became better known in the early sixteenth century, and then the Pacific in the late eighteenth, both Dominica and Tahiti formed one pole of a set of persistent dualisms, contrasted with other islands and their peoples in terms of degrees of savagery: Dominica, as the home of the cannibalistic Caribs, was contrasted with Hispaniola on which were found the supposedly peaceful and welcoming Arawaks, while Tahiti, as the home of the peaceful and welcoming Polynesians, was contrasted with New Caledonia and Fiji on which were found the man-eating Melanesians. In broad terms, therefore, a distinction was made in both cases between two groups—Arawak and Carib in the Caribbean, Polynesian and Melanesian in the Pacific. Arawaks and Polynesians were seen as being organized into relatively advanced, hierarchical societies; as responding relatively positively to Europeans; and, in physical appearance, as being closer to Europeans. Caribs and Melanesians supposedly had a less advanced, more egalitarian social system, were usually hostile to Europeans, and were seen as physically less attractive. In addition, they often indulged in practices that marked them as archetypally savage, especially cannibalism.

Although Dominica and Tahiti therefore seem to occupy different time frames and different poles of the established dualisms, there are general analogies between the ways in which the ethnic and cultural regions of the Caribbean and the Pacific were approached after contact, the ways in which those divisions became the basis for anthropological and historical work during the twentieth century, and the ways in which the divisions were undermined in the final quarter of the twentieth century—even though they arguably continue to exert enormous influence, perhaps more so in the Caribbean than in the Pacific.

But even the apparent differences in time frame are misleading in a number of respects. For a start, strong resistance by the indigenous Carib community limited European contact with Dominica until the seventeenth century, with major settlement only following much later, by which time the European view of the native population had begun to change. So the late eighteenth century was really the first occasion when British attention was actually directed at Dominica, the same moment that Tahiti hove into view. Both Dominica and Tahiti therefore played their parts in what has come to be considered as the first world war—the long and bitter conflict between Britain and France that followed the major American realignments of the Treaty of Paris in 1763. Dominica was a pawn in that particular game, passing into British hands. After 1763 both Britain and France immediately turned their attention to the Pacific, and both landed expeditions on Tahiti within months of each other in the late

1760s, led by Samuel Wallis for Britain and by Louis de Bougainville for France. During these years, contacts between Polynesian and Caribbean islands were certainly plentiful, exemplified by Captain William Bligh's successful breadfruit expedition in the *Providence* in the early 1790s. But many sailors, both French and English, who landed on Tahiti would have had previous experience of Dominica. Bligh and Fletcher Christian, for example, had been in the West Indies twice on the *Britannia* in the years before the *Bounty*. In reverse, Bougainville's naval career came to an effective end at the Battle of the Saints (the Battle of Dominica as the French call it) in 1782. In general terms, then, the late eighteenth century saw the establishment of a physical frame of reference that served to bring Tahiti and Dominica closer together. An intellectual frame would follow.

2

The dominant notes in early Spanish accounts of the Caribbean had been on the beauty of the climate and the landscape, on the paucity of indigenous culture, and on the bodies of the natives—well-formed and handsome, and in color neither black nor white but, according to Columbus, the color of the Canarian islanders, usually described as olive or copper.[2] There is an Edenic note in these early descriptions, conveyed by the use of terminology drawn from classical accounts of the golden age: "They lyve without any certayne dwelling places," wrote Peter Martyr about the native Caribbeans, "and without tyllage or culturying of the grounde, as wee reade of them whiche in olde tyme lyved in the golden age."[3] Climate was key here, since it supposedly stimulated the natural bounty of the soil and obviated the need for labor. In turn the absence of labor enabled the inhabitants of the Caribbean islands "to live at libertie, in play and pastime."[4]

But the ethnographic splitting between different groups soon appeared, in fact within Columbus's own journal. Some ideological dualisms are contrastive between self and other; this one, however, was triadic: in other words it depended on the supposedly neutral position of the European observer with respect to a supposedly observed division within the indigenous population, although the establishment of that division enabled a series of antagonisms and identifications that facilitated European entry into the politics of the Caribbean world.[5]

In broad terms, the European entry into the Pacific operated in the same way. The earliest descriptions of Tahiti, those by Philibert Commerson, by Louis de Bougainville, and in the anonymous *Relation de la découverte,* repeated the classical golden age vocabulary, confirming the Pacific islands as true islands of the west, Hesperidean in their climate and attributes. Commerson's early account of Tahiti struck exactly the same note as Peter Martyr's of the Caribbean: a place "where men live without vices, without prejudices, without want, without dissent . . . nourished by the fruits of the soil which is fertile without cultivation."[6] As the anonymous English poem, *Otaheite,* of 1774 has it: "No annual Toil the foodful Plants demand, / But unrenew'd to rising Ages stand."[7] Or as George Hamilton from the *Pandora* put it in 1791: "And what poetic fiction has painted of Eden, or Arcadia, is here realized, where the earth without tillage produces both food and cloathing."[8]

But the first classic statements of Pacific dualism are already found in the early accounts, first in Charles de Brosses's *Histoire des navigations aux terres australes* (1756) and then in Johann Reinhold Forster's "Remarks on the Human Species in the South-Sea Isles." Forster notes that "two great varieties of people" could be observed in the South Seas: "The one more fair, well-limbed, athletic, of a fine size, and a kind benevolent temper; the other blacker, the hair just beginning to become crisp, the body more slender and low, and their temper, if possible more brisk, though somewhat mistrustful."[9]

In both Caribbean and Pacific cases, then, the ethnographic division is based on what might be called temperament or character: one group is innately peaceful, the other innately warlike. In both cases, supposedly essential traits are "as much moral as physical," with the first being read off from the second: what Nicholas Thomas calls a "happy correspondence" between the advancement of different peoples and their perceived sense of appropriate behavior toward foreigners.[10] Although offered as ethnographic descriptions, these divisions in fact represent differential indigenous responses to European presence, and frequently they act as self-fulfilling from the describer's perspective. As such, they say more about European discursive practices than they do about the indigenous cultures of the Caribbean and the Pacific. In both cases discursive developments seem to have taken remarkably similar courses. Particular islands emerged as problem cases. Dominica was supposedly the archetypal Carib island but, given that the colonial description involved a narrative of Carib hostility and conquest and rape by Carib men, all second-generation Caribs must have been half-Arawak, and indeed the Caribs spoke a language that proved to be Arawakan. Correspondingly, Tahiti was sometimes seen itself as having two races, which led Dumont d'Urville, the French explorer, to concoct a conquest narrative in which Melanesians were conquered by Polynesians, but with their interbred remnants enduring in certain lower classes in parts of Polynesia.[11] The two conquest narratives needed to "explain" anomalies were mirror images of each other, but both

FIGURE 5.1 *Globe Divided into Frigid, Temperate and Torrid Zones,* from A. A. T. Macrobius, *Macrobii Aurelii Theodosii viri consularis in Somnium Scipionis libri duo, et septem eiusdem libri Saturnaliorum,* 1521 (James Ford Bell Library, University of Minnesota, Minneapolis)

served only to perpetuate and exacerbate those anomalies and, ultimately, to undermine the dualisms they had been supposed to help maintain.

In some obvious ways the Caribbean (and America more generally) provided a vocabulary that was used to describe the indigenous population of the Pacific. The term "Indian" continued its extremely slow circumnavigation of the globe; and the broad division between black and yellow/red peoples observed from an invisible neutrality was clearly available for the Pacific. However, the ideological imperative behind the "observation" of tropical ethnographic dualisms put pressure on the long-established climatic elements of tropicality. In general terms, Europe inherited the classical idea of an ideal temperate zone located between the frigid and the torrid zones in which savagery could be explained through oppression by cold and heat, respectively, the Scythians and Ethiopians offering the usual examples (fig. 5.1; see also chap. 11). Civilization was associated only with the temperate zone.[12] In the sixteenth century, Jean Bodin has similar divisions, with the temperate zone located between 30° and 60° north, the most temperate zone between 40° and 50°, a very French definition that excludes all of Britain apart from a few miles of Cornish coast.[13] For a classical geographer like Claudius Ptolemy the mid color of three would have been associated with the temperate clime because white was the color of the frigid Scythians. This was not a satisfactory link for northern Europeans, particularly the British, but also the Dutch, the Prussians, and the Swedes—all of whom were located north of 50°, and who, toward the end of the eighteenth century, began to redefine whiteness as the implicit standard. So to the extent to which skin color formed an element within the ethnographic dualism—which it did from the start in the Pacific—then the explanation of the existence of two colors within one zone needed to come from migration.

3

As far as the Caribbean is concerned, historical circumstances had changed a good deal by the end of the eighteenth century, most obviously through the genocide of the indigenous population and the transportation to the region of several million Africans. By the last decade of the eighteenth century, the region was in complete turmoil: Dominica was in the midst of a Maroon war, and, even more threateningly for British interests, St. Vincent—the other island with a significant Carib presence—was seeing an alliance between Carib insurgents and French revolutionaries. In various different ways, these changes resulted in the strengthening of the original dualisms but it also saw, through the Africanization of the Caribs, a strengthening of the Caribbean/Pacific analogy in ways that served to bring Tahiti and Dominica closer together.

The argument through migration actually finds its clearest form a little later, in the middle of the nineteenth century, when Carib origins were discussed in one of the institutional forums where the discipline of anthropology found its eventual shape, the Ethnological Society of London. Here James Kennedy rewrote the history of the early Caribbean to suggest that Columbus encountered "a widely-diffused tribe of dark colour and peculiar ferocity . . . designated Caribs or Cannibals," who were too intractable to submit to any intercourse with the Spaniards. Darkness of skin color, ferocity, intractability—by the 1840s the associations were inescapable, and so an African origin was posited for the Caribs.[14]

This Africanization of the Caribs may well have been influenced by the more explicit color dimension to the Pacific dualism: John Hawkesworth had referred to the Melanesians as "negroes" in his redaction of Cook's first Pacific voyage; Forster had referred to them as "blacker" than Polynesians. Wars against the Caribs in St. Vincent during the 1770s had certainly darkened them in British eyes, especially as they were by now well-mixed with escaped African slaves, according to the British story. Eventually—though not until the heyday of migration theory in 1949—someone suggested the irresistible conclusion that the Caribs actually descended from itinerant Melanesians and the Arawaks from Polynesians, thus "explaining" the remarkable analogy between European classifications of the indigenous cultures of the two regions.[15]

But if migration theory eventually worked to keep Caribs and Tahitians apart as opposite poles of that ethnographic dualism, one black, the

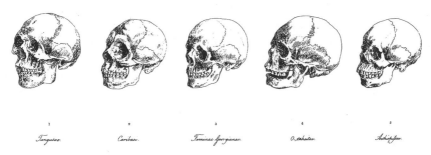

Plate iv

1   2   3   4   5

Tangusae.   *Caribaei.*   *Feminae Georgianae.*   *O-taheitae.*   *Aethiopissae.*

FIGURE 5.2 Johann Friedrich Blumenbach, *Tangusae—Caribaei—Feminae Georgianae—O-taheitae—Aethiopissae,* from Thomas Bendyshe, ed. and trans., *The Anthropological Treatises of Johann Friedrich Blumenbach* (London: Anthropological Society, 1865)

other yellow, another late eighteenth-century theory had found a new framework in which similarities and differences could be given equal weight. The third edition of Johann Friedrich Blumenbach's *De generis humani varietate nativa* was published in 1795, at the very moment of the outbreak of the war in St. Vincent that would lead to the final military defeat of the Caribs. For Blumenbach, there were five principal varieties of the human species: the middle, or Caucasian, variety; two extremes, Mongolian and Ethiopic; and two intermediate varieties, the Malay and the American (fig. 5.2). Five skulls took pride of place in the engravings illustrating this treatise, one for each variety: the American (second from the left) was represented by the skull of a Carib chief from St. Vincent and the Malay (second from the right) by a Tahitian skull.[16] For reasons that will shortly become apparent, Blumenbach may have preferred to have had a Carib skull from Dominica, but the island was in turmoil in the early 1790s, and St. Vincent had the great benefit of being the home of one of Joseph Banks's agents, Alexander Anderson, who ran the botanical garden that would eventually receive Bligh's breadfruit. Indeed, if it had not been for the mutiny, the *Bounty* would have unloaded the breadfruit and taken back to Europe the skull that Anderson had dug up at dead of night from a Carib burial site, and that Banks eventually sent on to Blumenbach.[17] The Tahitian skull was also a gift from Banks, brought back to England by Bligh on the *Providence*.[18] Blumenbach's essentialist approach assumed that within human variation there has to be an original template, which he predictably thought was the Caucasian, with "white" therefore the original human skin color, represented by what he called "the pure white skin of the German lady."[19] Variation depended partly on climate, so the "distant" Asian and African had traveled furthest from the Caucasian origin, toward frigid Mongolia and torrid Africa, respectively, whereas the two intermediate kinds, represented by the Carib and the Tahitian, belonged to the northern and southern tropics, respectively.[20] So tropicality now also began to have a racial dimension that separated it from the blackness of Africa. Caribbean and Pacific natives started to become—as they have remained—"brown" peoples, but this brownness was no longer the ideal midpoint between too cold and too hot, too white and too black, as it had been for the Greeks: it now marked a midpoint in the falling away from an ideal whiteness. This was the crucial turning point in the development of the racial ideology that forms the backcloth to all late-imperial writing.

But it was what had happened on St. Vincent that ultimately served to bring together Tahiti and Dominica—and their indigenous populations. Already by the second half of the seventeenth century, the remaining Carib population on Dominica had been too small to pose any significant military threat to European colonists, and the very earliest Columbian tropes were beginning to reappear. For example, at the beginning of the second volume of his *Histoire générale des Antilles,* written in 1667, Jean Baptiste Du Tertre—a French missionary in the Caribbean—set out to overturn two related falsehoods, one about the uninhabitability of the torrid zone, the other about the barbarity of its savage inhabitants. Just as the torrid zone is not "an awful wilderness" but, rather, "the purest, healthiest, & most temperate of all the atmospheres" (giving an interesting twist to the distinction between climatic zones by describing the tropical zone as more temperate than the temperate zone itself), so, similarly, according to Du Tertre, can the savage inhabitants of the tropics be described as not "barbarous, cruel, inhuman, without reason, [and] deformed" but, rather, "the most content, the happiest, the least depraved, the most sociable, the least deformed, and the least troubled by illness, of all the nations in the world."[21] It is Du Tertre who Rousseau references for the several mentions of the Caribs in his *Discourse on the Origins of Inequality.* So—at least within French ethnography and political philosophy—the Caribs had already been partially rehabilitated and may even, via Rousseau's reading of Du Tertre, have had an influence on the early French accounts of Tahitians, especially on Commerson's.

The major issue on St. Vincent, however, involved the question of the "Black Caribs," as they were called, a new ethnic group identified—or perhaps invented—by the British. These Black Caribs had joined French revolutionary forces to fight against the British in the 1770s and again in the 1790s.[22] The British argument was that these hybrid Black Caribs—part Carib, part African—were really Africans masquerading as Caribs, having stolen some Carib women. So the old dualism survived—or was reinvented—but now the Yellow Caribs, as they were newly called, a remnant left on Dominica, were allowed to assume the role of the good savage while the Black Caribs, seen as, for all intents and purposes, Africans, occupied the negative role. At this moment the indigenous inhabitants of Dominica and Tahiti finally came into alignment, and they did so because they both finally offered the most advantageous position ever open to indigenous peoples within colonial discourse: they were not black.

The key text here is Bryan Edwards's *The History . . . of the British Colonies in the West Indies,* first published in 1793, which was crucial for its realignment of British understanding of the Caribbean in the light of the ongoing French Revolution and the unfolding of events in St. Domingue. It is in Edwards's book that the Yellow Caribs could take on their new role as the Tahitians of the Caribbean, a move Edwards accomplishes by means of frequent references to Hawkesworth. This is just one example:

> Having . . . mentioned the natives of the South-sea Islands, I cannot but advert to the wonderful similarity observable in many respects, between our ill-fated West Indians and that placid people. The same frank and affectionate temper, the same chearful simplicity, gentleness and candour;—a behaviour, devoid of meanness and treachery, of cruelty and revenge, are apparent in the character of both:—and although placed as so great a distance from each other, and divided by the intervention of the American Continent, we may trace a resemblance even in many of their customs and institutions. . . . Placed alike in a happy medium, between savage life, properly so called, and the refinements of polished society, they are found equally exempt from the sordid corporeal distresses and sanguinary passions of the former state, and the artificial necessities, the restraints and solicitudes of the latter.[23]

As with Blumenbach, the triadic relationship is now stadial, with the Tahitians and Caribs sharing a position intermediate between savage life "properly so-called" and the civilized life of Europeans.

4

Although the persistence of this ethnographic dualism has bedeviled historical analysis of both Caribbean and Pacific cultures, one particular element within the discourse of tropicality is so pervasive that it pays no attention to dualisms: the trope of the spontaneous productivity of nature underlies European perceptions of indigenous indolence and absence of culture. This trope almost always draws its terminology from classical writing, as earlier quotations have suggested, but it turns the perception of the golden age against its inhabitants, who are ultimately seen as undeserving of its benefits. One of the founding comments about tropicality was made by Queen Isabella of Castile when told by Columbus that Caribbean trees have shallow roots on account of the high rainfall that makes the land so productive: "This land," Isabella said, "where the trees are not firmly rooted, must produce men of little truthfulness and less constancy."[24] And soon these ungrounded inhabitants were being described as intrinsically idle, drifting aimlessly over the surface of their lands—especially of course when Europeans wanted to make them labor.

Between the European discoveries of the Caribbean and the Pacific lies the development of a universalizing political philosophy that, in distinguishing between civilization and savagery, or sometimes even between the rational and the nonrational, also operates a distinction between the temperate and the tropical. John Locke gave the most powerful articulation of a labor theory of value that doubled as a test for distinguishing between the truly human and the less truly human, between those who had a right to own the earth and those who had forfeited that right by ignoring the law of nature that insisted that land should be "improved." In order to explain how indigenous peoples had fed themselves while not demonstrating full rationality, Locke had recourse to the Ovidian concept of the "spontaneous hand of nature," which had endowed the largely tropical lands of the Americas with a fertility that enabled their inhabitants to "labor" only by picking what nature had spontaneously provided for them. As George Sandys's contemporary translation put it: "The yet-free Earth did of her own accord / (Vntorne with ploughs) all sorts of fruit afford"; "of her own accord," translating Ovid's *sponte sua,* giving Locke his "spontaneous." The argument from spontaneity is necessary to explain how the "not fully rational" can manage to eat: it is this trope that excludes native American agriculture from consideration and even from recognition. In excluding native agriculture, it excludes native labor: the spontaneous hand of nature underlies native indolence.[25]

This trope of spontaneous nature, perhaps already inflected by Locke's labor theory of humanity, dominates early accounts of Tahiti. "Not," Joseph Banks admits, "that the trees grow here spontaneously but if a man should in the course of his life time plant 10 such trees, which if well done might take the labor of an hour or thereabouts, he would as compleatly fulfull his duty to his own as well as future generations as we natives of less temperate climates can do by toiling in the cold of winter to sew [*sic*] and in the heat of summer to reap the annual produce of our soil." Banks repeats Du Tertre's rhetorical willingness temporarily to relinquish the European association with temperance: for these purposes Europe is less temperate than Tahiti. But what turns the trope is the choice of classical tag: "*O fortunati nimium sua si bona norint* may most truly be applied to these people; benevolent nature has not only supplyd them with nescessaries but with abundance of superfluities." Any tropical natives hearing a quotation from the *Georgics* know that they are about to be accused of not working hard enough: "Could they but know their blessedness" as the Latin tag has it. "The great facility," Banks continues, "with which these people have always procurd the necessaries of life may very reasonably be thought to have originaly sunk them into a kind of indolence which has as it were benumbed their inventions."[26]

Although the ethnographic dualism haunts the historical anthropology of both the Caribbean and the Pacific, and to that extent their contemporary imagery, it is the trope of spontaneous nature that has passed seamlessly into the present. Only recently has scholarly attention been given to the sophisticated development of Caribbean tropical crops such as cassava and beans and maize in the centuries, indeed millennia, before European arrival; and in Tahiti, too, Dana Lepofsky's work has begun to uncover the extent of indigenous Tahitian cultivation, often—again repeating a Caribbean pattern—in arboricultural zones that Europeans did not even recognize as cultivated gardens.[27] The misrecognition of native labor as spontaneous bounty is the hallmark of contemporary tropical tourism and its associated writing, which will provide one final example of Tahiti and Dominica being brought into the same frame.

5

Paul Fussell begins his book *Abroad,* about literary traveling in the 1920s and 1930s, by discussing the way oranges came, in 1916, to symbolize a world away from the trenches of the Great War, one of the occasions when tropicality could come into its own, as temperance failed to live up to its name.[28] A generation of British writers who experienced that war subsequently made a career out of traveling and writing, often in the tropics or, at least, in places warmer than the fields of Flanders: Osbert Sitwell, D. H. Lawrence, Gerald Brenan, Robert Byron, Norman Douglas, Peter Fleming, Evelyn Waugh.

Waugh's elder brother, Alec, was the most assiduous of these travelers, taking advantage of the development of the ocean liners that allowed him in 1926 to go round the world for less than he would have paid to rent a London flat. His own retrospective account of these years stresses the perceived instability of the fabric of Western existence, a generational loss of faith that made it easy to wonder, he says, "whether the Polynesians had not built on more sound foundations. They lived by all accounts," he continues, "without wars and jealousies, without class distinctions, careless of their possessions, lovers of the sun. Surely it was worth going there to

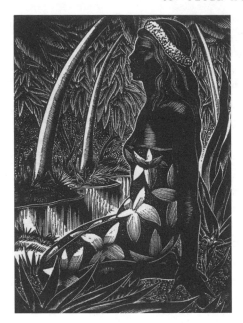

FIGURE 5.3 Frontispiece of Alec Waugh's *Hot Countries* (New York: Literary Guild, 1930), engraving by Lynd Ward

see?"[29] The "dark ladies" of Polynesia were an added attraction, a welcome antidote to the more independently minded metropolitan young women of the 1920s. What an earlier generation had seen as the degenerative features of tropical life were now, with disease supposedly taken out of the equation, embraced as its positives: inertia, alcohol, and sex. As Waugh himself says, he proceeded to search for his plots and characters "between Capricorn and Cancer": he became a specifically tropical writer.[30]

Waugh had his archetypal Tahitian experience: a brief passionate love affair with a Tahitian girl that he wrote up in fictional form as the affair of an acquaintance, before later admitting it as his own. But in literary terms the island had been written out, Waugh felt: there were no more stories to tell. "The South Seas . . . have been so written about and painted," he says. "Long before you get to them you know precisely what you are to find. There have been Maugham and Loti and Stevenson and Brooke. . . . And there has been Gauguin."[31]

En route for Tahiti on a French liner, Waugh had stopped briefly at Martinique. A year later, now actively looking for a subject, he thought of going to Martinique in order to write a comparison of the two French islands, which in fact he did in a 1930 book originally called *Coloured Countries* but issued in the United States as *Hot Countries*. At the end of the story of his Tahitian romance, Waugh had his protagonist reflect on the irony that his determination to settle in Tahiti had been foiled by the

FIGURE 5.4 Illustration from Alec Waugh's *Hot Countries* (New York: Literary Guild, 1930), 69, engraving by Lynd Ward

spirit of Tahiti itself, embodied by his Tahitian lover: "It was by her own loveliness, her own sweetness, her own gentleness, that Tahiti had been betrayed. . . . The fatal gift of beauty."[32] By contrast, "I found in the Caribbean the fresh material I needed," he wrote. "I was excited by the dramatic history of the area, I was moved by its beauty, I was fascinated by the West Indians themselves. They were so friendly, so willing, so fierce and so intractable. . . . Here were the stories I was looking for."[33] So here the old contrast is partly restored, now figured as a difference between the gentleness of Tahiti and the intractability of the West Indies, but with the latter proving more amenable to a writer in search of material—although Lynd Ward's stunning engravings for *Hot Countries* tend to undermine Waugh's contrast by conjuring up a singular tropical vision of bananas and palm trees and sexuality (figs. 5.3 and 5.4). Waugh became a regular visitor to the Caribbean and one of the most important interpreters of its history and politics during the twenty-five years before independence: his novel, *Island in the Sun* (1955) and the film made from it starring Harry Belafonte, is one of the best examples of the genre of late-imperial tropical fiction.

But Waugh also undermined his own contrast. The island that most fascinated him was, predictably, Dominica. "Of all the West Indian islands that I had visited," he wrote in 1948, "it was the one that I had liked the least; at the same time it was the one I was most anxious to see again . . . Dominica . . . has been called 'The Tahiti of the Caribbean.'"[34] Dominica clearly troubled Waugh, but he could not keep away from the place. He sensed defeatism beneath its desperate gaiety but came to find an alluring charm in that very acceptance of defeat. He described it as perfect in the grandeur of its mountains yet cursed with what he called—adopting the same phrase from Byron that he had used to define Tahiti—"the fatal gift" of beauty, the title he gave to his final novel, written in his seventies and set on Dominica.[35]

"The fatal gift" is a resonant phrase, recalling both the association of happiness with death in the old mythologies of the Western isles and the fatal gift of syphilis that Europeans brought to Tahiti—death introduced into paradise, as Diderot had intimated in his *Supplement to the Voyage of Bougainville* (1772).[36] The destruction resulting from European presence is transferred to the islands themselves, which then come to be seen as metonyms for their respective regions. Waugh's late-imperial contribution is to offer the phrase as a self-description of the European remnant that survives there, often victims of paranoia and alcoholism, the last white Creoles, who may have a desperate and rather touching commitment to the island of their dreams. Raymond Peronne, the central character of Waugh's *The Fatal Gift,* ends that novel in 1970—at a time when Dominica was in political turmoil—believing himself physically prevented from leaving the island because of a spell cast by an obeah woman jealous that he has deprived her of her lover—the sixteen-year-old girl whose affections they have shared: a classically tropical story. The white man is imprisoned after the British administration leaves, just as on the real island black Dominicans were gearing up to take the final step to full independence, the point at which their story draws apart from that of Tahiti.

Imagining the Tropical Colony: Henry Smeathman and the Termites of Sierra Leone

STARR DOUGLAS AND FELIX DRIVER

In January 1781, the naturalist Henry Smeathman presented a learned treatise to the Royal Society of London, under the title "Some Account of the Termites, Which Are Found in Africa and Other Hot Climates."[1] The treatise, subsequently published in the *Philosophical Transactions of the Royal Society,* has long been regarded by entomologists as the first detailed account of the life cycle and habits of tropical termites. Indeed, its author has been dubbed the "father of termitology."[2] Smeathman's paper was based on travels in Sierra Leone and in the West Indies during the 1770s, when he was sponsored by some of the leading naturalists and collectors of the day. By carefully cultivating his contacts within a variety of intersecting scientific, social, commercial, and political networks, in both the metropolitan worlds of natural history and the tropical regions through which he traveled, Smeathman effectively created a niche for himself within the rapidly developing field of entomology. In the process, he became an authority.[3]

During his stay in West Africa and the Caribbean, Smeathman collected thousands of entomological and botanical specimens on behalf of his various sponsors. As befitted a truly philosophical naturalist, he also recorded his observations on agrarian practices, domestic habits, trading networks, family structures, and social institutions. The significance of Smeathman's work in the context of the themes addressed in this book lies not simply in his role as a gatherer of knowledge about the tropical world but also in the way he conceived and framed this knowledge in both comparative and visual terms. In the course of his travels during the 1770s, Smeathman effectively followed the route of the triangular trade—from

England to West Africa, and thence to the Caribbean, before returning home in 1779—relying heavily on well-established trading networks spanning the Atlantic world. He used this experience to draw parallels and contrasts between different zones of the tropical world on either side of the Atlantic and experimented with the cultivation of plants and the control of pests. The business of collecting tropical nature during this period was frequently accompanied by the elaboration of ideologies of improvement—personal, moral, agrarian, and colonial—and in this respect Smeathman was an exemplary figure. His account of termite colonies published in *Philosophical Transactions* in 1781 was soon followed by his proposals for the establishment of a British colony in Sierra Leone. According to Smeathman, his plan of settlement—conceived in Africa, gestated in the West Indies and perfected in England—was the product of his comparative study of tropical nature in both its "rude" and its "cultivated" state. Smeathman's proposals played an instrumental role in the development of settlement schemes that came to fruition after his death in 1786, culminating eventually in the formal colonization of Sierra Leone.

Smeathman's writings on termite colonies and colonial settlement presented tropical nature as a terrain to be known and domesticated. His paper in *Philosophical Transactions* was accompanied by a series of striking illustrations, including a spectacular sketch of the habitations of the *Termes bellicosus* that has attained something of an iconic status among entomologists. The story of these images, their own natural history as it were, provides a particular focus for this chapter. Far more than passive accompaniments to the text, we shall argue, these images constituted particular ways of constructing tropical nature. Their combination in a single composite image rendered tropical landscape in simultaneously picturesque, topographic, and analytical terms. This fusion of modes of depiction usually considered distinct, if not contradictory, raises more general questions not only about composition and genre but also about the visual cultures of natural history.

A COMPOSITE IMAGE: SURFACE AND DEPTH

Smeathman's celebrated paper opens with a characteristic taxonomic gesture. Beginning with the names by which termites are known in West Africa, the Caribbean, Brazil, and "various other parts of the tropical regions" (including bug a bugs, white ants, scantz, and piercers), Smeathman immediately redescribes them in Linnaean terms, divided into five classes: *Termes bellicosus, mordax, atrox, destructor,* and *arborum.* It is the termitarium of *Termes bellicosus,* "about ten or twelve feet in perpendicular

FIGURE 6.1 Henry Smeathman, *The Termitary of "Termes bellicosus,"* engraved by James Basire (*Philosophical Transactions* 71 [1781]); Smeathman's original sketch is reproduced in plate 7

height above the common surface of the ground" that receives the most attention in what follows. Such a termitarium, pictured whole and in cross section, forms the centerpiece of one of the accompanying engravings, which also includes views of the smaller nest of *Termes arborum.* The remaining illustrations comprise cross sections of the royal chamber, nursery, mushrooms (food), and eggs of *Termes bellicosus;* turret nests of the *Termes atrox;* and heads and bodies of the various species of termites.

The most famous illustration accompanying Smeathman's paper is essentially a composite image, fusing elements of picturesque landscape representation, topographical mapping, and an analytical-anatomical gaze (fig. 6.1 and plate 7). Its depiction of topography is akin to the descriptive style of contemporaries such as Paul Sandby, who began his career as a military draftsman, rather than that of classical landscape artists (although some neoclassical elements were introduced by the engravers, as we shall see).[4] The sketch follows, at least in outline, rules of topographical composition, according to which the forms of nature were to be represented accurately and in proportion, based on careful observation and experiment. Such an empirical approach to landscape depiction is often said to have issued a fundamental challenge to neoclassicism. In the words of Barbara Maria Stafford, "The dictates of science urged a move away from the formulas of view hunting in the manner of Claude or Jacob von Ruisdael toward the naturalistic depiction of scenery that was worthy in itself." Topographical artists, she continues, "emulated the penetrating, retentive habit of observation instilled in the scientist. The process recapitulated the transference of aesthetic values from the artefact to the natural object."[5]

Elements of this mode of landscape depiction may be detected in Smeathman's approach to pictorial representation, as is apparent in the composition of the engraving in figure 6.1 and the original sketch in plate 7. Its exotic subject matter is presented in a conventional range of harmonizing colors—shades of brown ochre, green, and gray. The termitarium of *Termes bellicosus* forms the main focus, while figures in the background and cattle in the middle ground help, in part, to establish the scale of the termitarium. The light appears as if coming from the left, and Smeathman has shown alternating areas of light and shade. These "stabilising pictorial strategies" bear some similarity to those attributed to Sandby by art historian Charlotte Klonk: "Just as the sublime and the beautiful evoke eternal conditions, so topography stabilises the depicted scene, either by restricting itself to a limited range of harmonising colours independent of any characteristics that might stem from specific weather or time effects, or by strong compositional structures, which visually assert stability in the face of changing atmospheric conditions."[6] As in contemporary artistic depictions of remarkable rock formations and caves, the emphasis is on the accurate depiction of the "natural wonders of the world." However, as Klonk suggests, the result is less the application of a preexisting empiricism to landscape than a fusion of modes: "The topographical tradition, in which exactitude of delineation was paramount, was merging with the demands of the picturesque for roughness and variety."[7]

The composite nature of Smeathman's sketch becomes clear when one considers the form and significance of the cross-sectional views, which draw the viewer's eye into the labyrinthine interior of the termitarium. As with contemporary illustrations of grottoes, rock-cut temples, caves, and arches, this sketch draws attention to a curious and intricate structure. If the phenomenon is natural, however, the view is artificial: a depiction of a termitarium alongside a cross-sectional view of the interior that is not normally visible. The scientific eye unveils the invisible structure of the scene to the viewer, effectively combining the techniques of architectural and anatomical drawing. In this context, it is notable that Smeathman's descriptions of the internal structure of termitaries relied heavily, indeed exorbitantly, on an architectural vocabulary. As he reported, the dome of the termitarium of *Termes bellicosus* typically contained fortified structures, apartments, galleries, "gothic shaped arches," passages, pipes, bridges, scaffolds, and labyrinths. What is particularly striking here is the emphasis on the strength and complexity of these natural forms, which were less habitations than fortifications. The military metaphor was particularly apt in this context: like the string of European forts along the coast of West Africa, the termitaries were designed to repel invaders. Only through destruction could their intricate design be revealed: "These subterraneous passages or galleries are lined very thick with the same kind of clay of which the hill is composed, and ascend the inside of the outward shell in a spiral manner, and winding round the whole building up to the top intersect each other at different heights, opening either immediately into the dome in various places, and into the interior building, the new turrets, etc. or communicating thereto by other galleries of different bores or diameters, either circular or oval."[8]

Smeathman concedes that the task of describing these structures strained his vocabulary: as he put it, "they are so constructed upon so different a plan from anything else upon the earth, and so complicated, that I cannot find words equal to the task."[9] In a French edition of his treatise, published in 1786, he stages a still more fulsome apology to the reader for his frequent resort to metaphor, "to circumlocution, to paraphrasing, to comparisons that necessarily muddle the subject matter," though it is significant in this context that this apology is used to highlight the value of the illustrations: "I believe I have remedied this inconvenience, in part, by means of the plates that I have added to my work."[10] If the textual descriptions were often architectural, the graphic techniques owed much to anatomy—and, most important of all, both reflected a common impulse to render visible what lay beneath the surface. In this context, Stafford's description of Giovanni Battista Piranesi's technique of architectural drawing is particularly apt: "With scalpel-like wielding of the etcher's needle, he applied surgical procedures (learned, I suggest, from medical illustrations) to turn the still-living fabric of architecture inside out."[11] The practice of anatomical dissection reinforced the connections between the pursuit of medicine and natural history: in the case of insects, the first spectacular illustrations were made by anatomists using microscopes, including Robert Hooke, Jan Swammerdam, and Antoni van Leeuwenhoek. While the anatomical study of insects originated in the seventeenth century, the term "entomology" itself was in fact first coined by the Swiss naturalist Charles Bonnet in the 1760s, only a few years before Smeathman presented his paper to the Royal Society. The term contains within its Greek root the impulse to dissection and analysis that was essential to the definition of the field: the cutting up of bodies in order to render their internal form visible. By the time Smeathman's paper was

FIGURE 6.2 Henry Smeathman, *The Royal Chamber of "Termes bellicosus,"* engraved by James Basire (*Philosophical Transactions* 71 [1781])

read, microscopic analysis, anatomical dissection, and drawing had become an essential means of producing knowledge about insects.

Smeathman's cross section of the termitarium of *Termes bellicosus* itself has the appearance of contemporary anatomical drawings, as though the hills themselves were living organisms; it shows a comparison between the visible surface and the invisible interior and graphically depicts the different layers surrounding the royal chamber. The technique also enables a view of the termitarium and the termite body, at a series of scales. A dissection of the royal chamber thus provides the basis for a further more detailed sketch (fig. 6.2). In this image, the royal chamber is detached from its surroundings in order to represent transverse, longitudinal, and foreshortened sections with the termite queen, king, and laborers in attendance. Penetrating still further, Smeathman sketches the nursery itself, complete with its eggs, "mushrooms" and moldiness, "as just taken from the hill" (fig. 6.3). And finally, he provides a sketch of the individual heads and bodies of the various species of termites, showing queen, king, laborer, and soldier, both whole and with heads magnified (fig. 6.4). The topside and underside of the three orders are depicted; the queen is shown side-on, and the enlarged heads are shown separately. These different ways of placing the insect are designed not merely to highlight the most salient features, such as wings, mandibles, and legs (and, in the case of the queen, her immense bulk), but also to invite an analytical approach to the de-

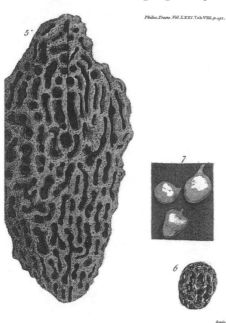

Philos.Trans.Vol.LXXI.Tab.VIII.p.191.

FIGURE 6.3 Henry Smeath-man, *Nursery, Mushrooms and Eggs of "Termes bellicosus,"* engraved by James Basire (*Philosophical Transactions* 71 [1781])

scription of bodies. The different species of termites, or parts thereof, are presented as if pinned to a flat surface, according to the manner of contemporary entomological illustration. These drawings were made on the basis of Smeathman's own microscopic analysis, though from evidence presented in his paper it appears that he was assisted by the anatomist John Hunter.[12]

In describing Smeathman's sketches as "composite images," we intend to draw attention to their multiple sources and effects. In some respects, we would argue, the combination of various different modes and scales of visual representation within a single image was not as exceptional as it might at first appear. It is tempting to draw a parallel here with some of Piranesi's archaeological illustrations, which have tellingly been characterized as "multi-informational images." These present what Susan Dixon calls a "dizzying juxtaposition of unlike imagery: a stark geometric section next to a palpable view—perhaps rendered with richly textured shadows."[13] Whereas Dixon presents this aspect of Piranesi's work as eclectic and disconcerting, her account of its reappearance in the literature of *voyages pittoresques* (the subject of Claudio Greppi's chapter in this volume) suggests it had resonance well beyond archaeological illustration. And in the work of naturalists such as Henry Smeathman, the combination of very different visual modes enabled, rather precisely, different views to be accommodated in the same frame.

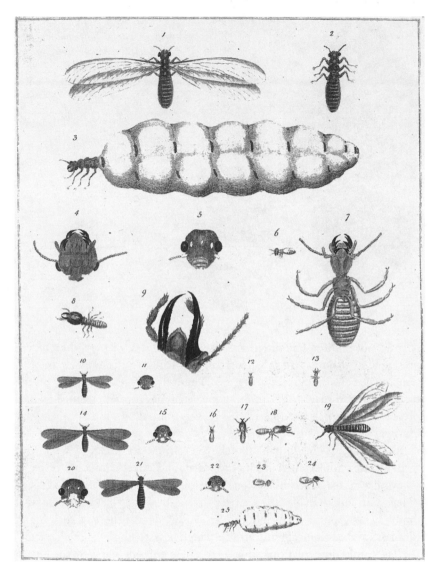

FIGURE 6.4 Henry Smeathman, *Heads and Bodies of Tropical Termites,* engraved by James Basire (*Philosophical Transactions* 71 [1781])

PICTURESQUE TROPICALITY

Smeathman was keen to establish the authenticity of his observations, and this required the adoption of a certain style of address and studied reference to numerous other works by naturalists and travelers who had set eyes on tropical termitaries. He explained the addition of footnotes to his treatise between presentation and publication in precisely these terms: "It is possible the accounts I have here communicated would not appear credible to many, without such vouchers and such corroborating testimony as I am fortunately able to produce, and are now before you."[14] His self-fashioning as a careful observer, familiar with the etiquette of respectable natural history, was entirely consistent with his recourse throughout the paper to the language of wonders and marvels. His opening descriptions of the termitaries set the tone: "The size and figure of their buildings have attracted the notice of many travelers and yet the world has not hitherto been furnished with a tolerable description of them, though their contrivance and execution scarce fall short of human ingenuity and prudence; but when we come to consider the wonderful economy of these insects, with the good order of their subterranean cities, they will appear foremost on the list of the wonders of creation, as most closely imitating mankind in provident industry and regular government."[15]

Smeathman's illustrations, like his text, had the dual function of rendering tropical nature simultaneously exotic and familiar.[16] The composition of the picturesque scene in plate 7, for example, reflects contemporary conventions in the depiction of agrarian landscape, though it also contains icons of tropicality, not least the termitaries themselves. On the right-hand side of the sketch is a herd of cattle, with the bull perched on a termite hill, as Smeathman had observed in Sierra Leone.[17] Judging from the shape of the horns and the humps on their backs, these cattle are probably zebu. Apart from their role in establishing a basic sense of perspective and scale, the inclusion of these animals adds an aspect of composed pastorality to the scene. Smeathman has here juxtaposed elements of tropical difference—the termite nests, palm trees, and a single representative native of Sierra Leone—with images of European endeavor and activity (notably the figures on the look-out in the far distance, also standing atop a termitarium), the whole set within a broadly picturesque scene. The tree framing the sketch on the left-hand side contains the nest of another species (*Termes arborum*), and its dissected counterpart is sketched to the right. The scene is artificial, as the two kinds of nests would not actually be observed in such close proximity: it is presented partly as an analytical construct, partly as a picturesque device, as the overarching bough of the tree provides a natural frame.

The balance between exoticism and familiarity was significantly affected by the reproduction of Smeathman's sketches through the process of copper-plate engraving. James Basire, the engraver responsible for translating Smeathman's sketches into print, also worked for the Society of Antiquaries and for publishers of voyages and travels.[18] The use of visual images to accompany papers published in *Philosophical Transactions* was becoming increasingly common in the second half of the eighteenth century. According to Bernard Smith, "Some of the engravings which appeared in the *Transactions* during this period possessed an arresting visual appeal quite apart from what they were intended to illustrate."[19] Given the nature and cost of the process of engraving, and the skills it required, the engraver had considerable influence on the final product. And as many art historians have shown, there were often significant discrepancies between various elements of the sketch and engraving.[20] This could be regarded, negatively, as the product of technical limitations: as Rudwick puts it in the context of geological illustration, "expression of subtleties of texture and shading had to be translated by the engraver on to the copper plate by means of visual conventions of hatching and cross-hatching."[21] But it could also involve the enhancement of detail and, notably, contrast. Comparing figure 6.1 with the original sketch in plate 7, for example, we can see that Basire made some minor modifications to the cattle on the right, so that the horns now resembled a Kerry and the shape, a Friesian. He also added more density and detail to the foliage overhanging the termitarium. While Smeathman's tree is not readily identifiable, in Basire's engraving the tree is recognizable from its leaves as an oak.[22]

There were other obvious differences between sketch and engraving, including a sharpening of contrast, the introduction of numbers and letters referring to the text, and (in the case of the French edition) a reversal of the orientation of the image. One of the most significant changes to detail is visible in the body of the native assistant, who in the engraved image acquires European features depicted more formally in the neoclassical style. In both sketch and engraving, the image invites a narrative reading, as the assistant stands holding the hoe that, we are led to presume, has been used to expose the interior of the termitarium. The evidence concerning Smeathman's actual fieldwork in Sierra Leone (predominantly within his own journal and letters) suggests that he relied heavily on assistants and intermediaries: indeed, without them, his work would have been simply impossible.[23] In the English edition, the engraver follows Smeathman's sketch in feminizing the assistant, depicted in passive mode, leaving his labor implied rather than shown. In the French version, however, the illustrations include an additional image of an African hacking into an intact termitarium, his muscles flexed in the process of destruction, reiterating in visual

FIGURE 6.5 Henry Smeathman, *Breaking Open the Termitary* (Smeathman, *Mémoire pour servir à l'histoire de quelques insectes, connus sous les noms de termès, ou fourmis blanches* [1786])

terms the notion of the termitarium as fortification (fig. 6.5). In both engraved versions, the figure is clearly idealized, if not as "Grecianized" as other images of non-European peoples, as for example in the celebrated examples of images of Pacific islanders analyzed by Bernard Smith and others.[24] While the transformation is not so dramatic in the case of Smeathman's images, it is readily apparent, and—as in the case of other elements, including the vegetation and animals—the effect is to adorn the scene with the familiar icons of the classical pastoral scene.

There is one further aspect to the scene that needs to be noted: in the background to figure 6.1 is a group of five men in European dress, standing atop yet another termite mound, looking through a telescope at a distant object. This vignette contains another narrative clue: in the accompanying text, Smeathman reports that "I have been with four men on the top of one of these hillocks. Whenever word was brought us of a vessel in sight, we immediately ran to some Bugga Bug hill, as they are called, and clambered up to get a good view."[25] The European eye surveying the horizon is a familiar image, gesturing to the ability of science and technology to reach beyond the local and, with the help of the text, reminding the reader of the connections between this tropical location and the wider world of circulation and trade. Here, as elsewhere, the trope could also be read rather differently, as an expression of the fragility of such connections across the seas. While in Sierra Leone, Smeathman lived and worked mostly in the Banana Islands, which were then under the control of various African-European traders, most notably, James Cleveland. These traders played a key role in the development of the Atlantic slave-trading system, and their influence with both Africans and Europeans was vital to everything Smeathman did in the field. The proximity of these figures and the assistant within the space of the image in figure 6.1 serves to highlight the differences between them, leaving no doubt which group is doing the surveying and which is the surveyed. The termitaries them-selves occupy the foreground, the perspective and cross-sectional views accentuating the sense of the alterity of tropical nature; the African assis-tant and the distant surveyors meanwhile serve to assure the viewer of the close association between knowledge and control.[26]

This play between the familiar and the exotic reflected a broader pat-tern in the literature of travel and natural history during the second half of the eighteenth century. In the words of Barbara Maria Stafford, "The puzzling and the marvellous were reckoned to be amenable to explanation and hence to reproduction. By this standard of intellectual conquest, il-lustration and description must not fall short of the thing it resembles."[27] Smeathman's rendering of the termitarium of *Termes bellicosus* in the con-text of a picturesque topographic scene has the effect of lessening some-what the menace of the exotic, while asserting its status as a curiosity. Here the visual image seems to offer a way of representing tropical differ-ence that manages to both retain and contain its otherness. As Kay Dian Kriz puts it in a compelling discussion of Sloane's *Natural History of Ja-maica,* "If the curiosities represented become too thoroughly pacified in the process of visual and verbal representation, their capacity for arousing the wonder and desire of the reader will be diminished sharply. Too little pacification threatens to expose an Otherness that cannot be known and, even more worrying, cannot be physically contained."[28] The iconic sta-tus acquired by Smeathman's sketch would suggest that he managed to strike an effective balance between familiarity and strangeness, rendering the termitarium a proper object for the inquiries of natural historians.

COLLECTING THE EXOTIC

Contemporary interest in the collection and depiction of exotic speci-mens reflected a variety of impulses, scientific, commercial, social, and re-ligious—in this respect, Smeathman's expedition to Sierra Leone, with its multiple sources of patronage, was symptomatic of a broader pattern. While Joseph Banks provided the channel through which his treatise was eventually published in the pages of the Royal Society's *Philosophical Trans-actions,* Smeathman should not be regarded simply as a footsoldier in the expanding Banksian empire.[29] Prior to his voyage to Sierra Leone in 1771, Smeathman was already an active participant in the highly competitive world of insect collectors and entomologists. The sponsors and supporters

FIGURE 6.6 Type specimen of *Goliathus goliatus* (Drury), from William Hunter's insect collection, Hunterian Museum (Zoology), University of Glasgow; photograph by Geoff Hancock

of his expedition included (in addition to Banks himself), Dr. John Fothergill, an influential figure in the transatlantic network of Quaker naturalists; Moses Harris, author of *The Aurelian* (1766), a sumptuously illustrated book on British butterflies and moths; and the silversmith Dru Drury, a leading member of the Aurelian Society of London, a select body that promoted the collection and study of insects.[30] The first volume of Drury's *Illustrations of Natural History* (1770), devoted to exotic insects, featured a particularly contentious specimen that had been found in tropical West Africa a few years before: the so-called Goliath beetle. The story of this beetle is of relevance here as it indicates the significance attached to the possession and visual representation of exotic specimens within such communities of collectors. It also sheds light on the immediate context of Smeathman's voyage to Sierra Leone in 1771.

In 1766 a ship's surgeon came into possession of one of the largest insects in the world, subsequently named the Goliath beetle (fig. 6.6). According to the account given in Drury's *Illustrations of Natural History,* the beetle "was brought from Africa by Mr. Ogilvie, now Surgeon to his Majesty's ship, the *Renown,* being found floating dead in the River Gaboon, opposite Princes Island, near the Equinoctial line."[31] The colored plate that accompanies this description is signed by Moses Harris, and dated 1767, and it includes the Fabrician classification "*Goliathus goliatus* (Drury)."[32] However, the beetle actually belonged to Dr. William Hunter, the celebrated surgeon and anatomist, who had purchased it from Ogilvie. The specimen came to be engraved by Harris after Hunter had lent it to Emanuel Mendes da Costa, a fellow member of the Aurelian Society and also clerk to the Royal Society, who was preparing a publication on natural history.[33] The latter's subsequent imprisonment for embezzling the society's funds may have played some part in the dispute over the engraving, which eventually appeared in Drury's *Illustrations* with an attribution to Drury, much to the annoyance of Hunter.[34] This classification was made by the Linnaean-trained Danish entomologist Johann Fabricius, who had visited Britain in 1767 and examined insects in the collections of Banks, Drury, William and John Hunter, Fothergill, James Lee, and Thomas Pennant, among others.[35] Fabricius must have presumed the Goliath belonged to Drury, who did not correct that assumption. Hunter wrote to Costa in King's Bench Prison demanding an explanation, and in a long and groveling reply, Costa admitted selling the plate to Drury without Hunter's consent. The response from Hunter was acerbic: "Mr. Da Costa's owning it was wrong is enough. But it must remain so. Dr. Hunter chuses no further dealings. He thinks Mr. Drury likewise has behaved in a way that he should not have expected. But if they are pleased with themselves he has nothing to say."[36] The letter is dated 10 January 1771, ten months before Smeathman set sail for Sierra Leone, appropriately enough, on a vessel called the *Fly*.

The dispute over the *Goliathus* specimen gives some indication of the intense rivalry between collectors over the acquisition and identification of exotic specimens, especially so-called type specimens, the first known example of a new species against which all subsequent examples were measured. Type specimens were eagerly sought after: they added to the prestige of the collector as well as the reputation of the collection, and if figured in an illustrated book, they added value to the book. This in itself placed a premium on the collection of ever more exotic insects, hitherto unknown to collectors in Europe and North America. While Harris's *Aurelian* was devoted to insects found in Britain, Drury's *Illustrations of Natural History* depicted mainly exotic specimens collected by ships' captains, traders, and others, many of them acquired in tropical regions. By the late 1760s, Drury had built up a global network of correspondents and collectors, from Africa to the Caribbean and from the China coast to the Americas.[37] In 1766, the year of Goliathus's discovery, he prepared instructions for these correspondents on the collection and preservation of specimens, offering sixpence for every specimen received safely.[38] While Drury was here following a recognized tradition in issuing such advice, what was notable about these instructions was that they were predominantly aimed at collectors traveling overseas, notably in the tropics. He was particularly concerned to ensure the safe preservation and transportation of specimens: collecting at a distance was a costly business.

In 1767, the year that Harris made the plate of the Goliath specimen, he also sketched a tarantula from Sierra Leone.[39] The literature of voyages and travels had long included reports of dangerous insects in tropical Africa, notably spiders and scorpions, though relatively little was known about them. In his narrative on the Guinea coast, for example, Jean Barbot reported that "some of the spiders seen here are of an astonishing and horrifying size," and he also described locusts, grasshoppers, scorpions, glowworms, "great black flies," gnats, and "large ants with yellow wings" (probably termites).[40] William Smith's *A New Voyage to Guinea* (1744) had also described the different ants (including termites, which he called "white ants"), scorpions, cockroaches, and "monstrous large spiders."[41] As this suggests, the emphasis was often on the sheer size of the insects, though a beetle as large as the Goliath had never previously been heard of: the naming of the specimen also reflected associations between size and power. It is not difficult to imagine the excitement that its appearance caused among the entomological fraternity in London. With Costa in prison, and unable to publish his own book, the specimen took pride of place in the first volume of Drury's *Illustrations,* privately published in 1770. In this volume, there were only two other specimens from Sierra Leone (both locusts), and most of the images were of insects collected in the Caribbean and the Americas. However, the third volume, published twelve years later, contained a large number of specimens from Sierra Leone, many attributed to Henry Smeathman.[42] The decision of Drury and other collectors to sponsor Smeathman's expedition to West Africa in 1771 must be seen in the wake of the *Goliathus* affair and their drive to acquire rare and exotic specimens. According to Smeathman himself, Sierra Leone had been selected "as a country the least known to Europeans, and the most likely to afford a variety of new, curious, and valuable specimens in the three kingdoms of Nature."[43]

COMPARATIVE NATURAL HISTORY: CIRCULATION AND COLONIZATION

Henry Smeathman spent roughly four years in Sierra Leone, during which time he kept journals and corresponded regularly with his sponsors. The observation and collection of exotic specimens was only one part of the naturalist's task, which generally could be characterized as a sort of territorial surveillance, involving a variety of geographical, meteorological,

botanical, zoological, and ethnographic observations. Smeathman's interactions with European traders and African-European settlers along the
West African coast suggest that he was, to a large extent, dependent on the
support of locals, as well as the patronage of metropolitan collectors. It
seems that toward the end of this period, he agreed to act as an agent for
a Liverpool merchant, whose ships he planned to use for the transportation of his collections.[44] It may have been on one of this trader's vessels
that Smeathman traveled in 1775 to the West Indies, where he was to remain for a further four years. While relatively little documentation survives for this period of Smeathman's life, it seems that his plans to return
immediately to England may have been interrupted by illness. His decision to remain was also influenced by the large rewards being offered in
Tobago, Grenada, Barbados, and other islands for the successful eradication of "cane ants" (also known as "fire ants," because of their bite) that
were seriously disrupting plantation agriculture. Whatever his motivation,
this period spent in the Caribbean was subsequently to be represented by
Smeathman as a decisive moment in the development of his thinking—
not only about termites and other insects, but about the possibility of
agriculture and settlement in the tropics. In a letter to John Lettsom, written in 1782, he drew together his experiences of tropical travel in terms
calculated to draw attention to his credentials as a comparative natural
historian:

> After a residence of about four years in Africa, I embarked with my collec-
> tion for Europe, by way of the West Indies; but being very ill on my arrival
> in Tobago, I determined to stay there, rather than meet the winter's winds,
> which the ships from thence, at that season, must necessarily encounter. I
> had seen the equinoctial lands in a state of nature, and was curious to mark
> the appearance of them in high cultivation. . . . My stay in the West Indies
> furnished opportunities of corroborating and improving the observations I
> made in Africa. There I became acquainted with tropical agriculture and
> manufactures, and much to my satisfaction.[45]

The language of "improvement" here had a variety of functions: as well
as accentuating Smeathman's self-fashioning as a gentlemanly natural historian, it also reflected the worldly aspirations of enlightenment science.
In the present context, moreover, it highlights what Smeathman regarded
as the virtues of the comparative view: having observed tropical nature "in
a state of nature," he was able to represent his Caribbean sojourn as spent
in the study of its nonidentical twin, "in high cultivation." This formulation enabled him to assert simultaneously the prodigiousness of tropical
nature and the superiority of European technology. He was thus sharply
critical of indigenous methods of rice cultivation in Sierra Leone (which,

like most travelers, he simply failed to comprehend) and became convinced of the possibility of establishing plantations in West Africa. (See also Peter Hulme's discussion of indigenous cultivation of tropical crops in the Caribbean and the Pacific in chap. 5.) With his experience across the Atlantic tropical world, he was able to make comparisons and contrasts—not only in the case of termites and polyps but also with varieties of rice and people.[46] His travels during the 1770s had, in fact, involved him in tracing the route of the triangular trade—from England to West Africa, then to the Caribbean and back. In Sierra Leone, he relied heavily on the knowledge and support of traders in the transatlantic economy, including those who made their fortunes out of slavery. Following his return to England in 1779, he became embroiled in a different sort of transatlantic network dedicated to the extinction of the slave trade.

Shortly before his death, Smeathman published a "Plan of Settlement," in support of a scheme to establish a colony of free blacks in Sierra Leone under the auspices of philanthropists and abolitionists.[47] While the plan was designed to promote the work of Jonas Hanway's Committee for Relieving the Black Poor in 1786, its vision of a well-ordered industrious community owed much to Smeathman's own philosophy of (self-) improvement. As he explained in a letter written to the Quaker physician Dr. Thomas Knowles in 1783:

> I conceived this project in Africa, where an industrious cultivation of the soil, with various excursions to different parts of the country, made me well acquainted with the genius, customs, and jurisprudence of the inhabitants, and the state of their agriculture, trade and arts. My stay in the West Indies was in the first instance entirely with a view of informing myself of the methods of cultivating tropical produce, etc. previous to my return to Africa. I accomplished my intention, and have since, by a constant attention to various branches of philosophy and useful arts, qualified myself still further.[48]

Smeathman exploited his experience as a naturalist in West Africa and the Caribbean to make the case for a British-led settlement in Sierra Leone. The possibility of introducing new crops and new systems of cultivation was justified on comparative grounds, chiming with broadly Banksian assumptions about the transfer of crops across the intertropical zone:

> [Sierra Leone] lays under climates, which in the other continents produce the richest materials of commerce, and its products are actually similar. . . . But as the ginseng is found both in North America and the northern parts of Asia . . . so in all probability there are few of the riches of the eastern or western hemispheres, which may not be found in this middle region, or at least commodities of equal value and importance.[49]

Seeds and crops were not the only things that could be transplanted, as Smeathman knew only too well from his direct observation of the up-rooting of Africans and their transportation across the Atlantic. In his let-ter to Dr. Knowles, Smeathman made great play of the philanthropic in-tentions of his scheme, which were further accentuated in his published plan. He envisaged a community made up of both black and white colonists from various parts of the world. The whites, he proposed, would be chiefly tradesmen, "carpenters, joiners, coopers, smiths, rope-makers, sail-makers, weavers, taylors, masons, gardeners, men bred on West India plantations, viz. planters, distillers, etc." The black population, in the ma-jority, were to be drawn from West Africa, the Caribbean, and England itself. "There are," he noted in 1783, "now in this country, hundreds, and many of them persons of character, possessed of a little property, who un-der the sanction of a respectable company of Quakers, and the prospect of an independent settlement, would gladly engage." Moreover, there were "vast numbers of *people of colour* in the West Indies, who though *free,* as we are pleased to term it, in those islands, labour under such intolerable op-pression and insults, that they would almost to a man unite themselves to such a community." Further supplies of black labor could be obtained in the Canaries and elsewhere in West Africa, especially if humanitarian sen-sibilities about the trade in slaves could be got round: "If it might be per-mitted to purchase a few slaves at Senegal as well as at Goree and Gambia, it would be political: many of the slaves from these parts have as just a sense of the value of liberty, as either Britons or Americans, and are brave and ingenious men."[50]

With more than an eye to his Quaker readers, Smeathman emphasized that the government of the colony would be conducted on enlightened principles: the land would be acquired through negotiation with local rulers, and the colony would be run according to "a regular Code of Laws." A division of labor would operate, with black Africans providing the manual labor, whites the management and skilled labor, and Smeath-man himself the expertise necessary to develop the resources of the colony. In his paper on termite colonies, Smeathman had particularly em-phasized the "provident industry" and "regular government" of the ter-mites, and he had described the different orders of insect in the language of political economy: laborers, soldiers, nobility (the winged insects), and queens. He had also attributed human qualities to the termites—ingenu-ity, industry, prudence, sagacity, good order—the same qualities that now appeared in his proposals for the human settlement of Sierra Leone. This "foundation of a commonwealth . . . would enjoy every degree of civil and religious liberty, be a sanctuary for that insulted and oppressed set of men (the people of color) and gradually abolish the unnatural trade in the human species on the Coasts of Africa." The colonists would grow mainly rice and cotton according to European rather than indigenous methods of cultivation: there were "large, fertile and unoccupied tracts of land," waiting to be cultivated by a more "industrious and commercial people." Forests would be cleared, to be replaced by plantations; and more colonists would supply additional labor. According to Smeathman, the infant colony would eventually incorporate neighboring territories into its orbit, as if it were the *bellicosus* of the human world:

> In short, if a community of two or three hundred persons were to be associated on such principles as constitute the prosperity of civilized nations, such are the fertility of the climate, the value of its products, and the evident advantages, of such an establishment, that it must, according to the common course of things, with the blessing of the Almighty, encrease [sic] in wealth and numbers with a rapidity beyond all example; and in all probability, extend its saving influence in the short period of thirty to forty years, wider than even *American Independence*.[51]

Of course the subsequent history of settlement in Sierra Leone was to be a rather different affair. The new settlement eventually established (after Smeathman's death) on the southern shore of the Sierra Leone river in 1787 was short-lived: the rains washed away the topsoil, fever ravaged the community, and the settlers soon began trading their muskets and other goods with the local Temne for rice. But the alliance between philanthropy and natural history reflected in Smeathman's original plan of settlement was to be revived in subsequent projects. The failure of the initial plan was followed by a more organized venture (under the auspices of the Sierra Leone Company), composed of settlers from a variety of locations, including freed slaves from Nova Scotia who had fought on the British side in the American War of Independence. In 1792, Dr. Thomas Winterbottom took up the post as company physician, and his *Account of the Native Africans of Sierra Leone* (published in 1803) is faithful to the enlightenment model of natural history as territorial surveillance, presenting an overview of landscape, climate, agriculture, diet, habitations, employment, trade, religion, and medicine. Winterbottom's narrative presented the settlement at Sierra Leone as an ordered community, its natural resources plentiful and its people worthy of the civilizing effects of education and good government—in other words, improvement. In this context, Smeathman's treatise on termites—reprinted in an appendix to Winterbottom's *Account*—served to confirm the potential of human mastery over tropical nature and to enable its application in practice. Just as the colonies of termites, once so alien and threatening, had been brought to order by patient observation and experiment, so too would its human counterpart in the tropics.[52] Five years later, Sierra Leone became Britain's first Crown colony in tropical Africa.

VISIBLE NATURE, INVISIBLE NATURALIST?

In this chapter, we have considered the contribution of a little-remembered naturalist to the visualization of the tropical world. The role of such figures in the production of natural knowledge has often been overshadowed by the work of the philosophers and princes of European natural history—Linnaeus, Banks, Buffon, Humboldt. Here we have used Smeathman's work on tropical termitaria to highlight the multiple sources and composite forms of knowledge about tropical nature. We have also drawn attention to the ways in which such relatively humble collectors of exotic specimens could actively exploit the various networks in which they operated. The success of Smeathman's career as a natural historian, judged in his own terms, owed much to his careful cultivation of relationships with many of the key figures in the world of collecting and natural history, within and beyond British shores. While his contributions are not much remembered today, his name remains a silent presence in the taxonomic schema of botanists and entomologists, courtesy of Banks and Fabricius, who attached it to a species of passion fruit tree in West Africa (*Smeathmannia*) and a humble moth (*Pyralis Smeathmanniana*).[53]

By cultivating relationships with learned natural philosophers and influential collectors within the metropolitan community, Smeathman was able to gain access to a variety of social, political, technical, and financial resources. Some of these connections are written into the very fabric of his treatise in *Philosophical Transactions,* in its mode of address, its learned style, its footnotes, its metaphors, and its allusions. Less visible are the networks of support on which Smeathman relied during his travels within the tropical world, notably the well-established infrastructure of African-European society and trade on the Guinea coast. Far from venturing single-handedly into uncharted regions, Smeathman was utterly dependent on existing structures of patronage and influence, built up over the previous decades, dominated by the Atlantic trading system. With local assistance, direct and indirect, he collected, classified, sketched, preserved, and transported specimens of tropical nature to his various sponsors in England and to Linnaeus in Sweden. And in the process, as we have said, he himself became an authority.

While Smeathman's treatise does refer to the particulars of his own experience in the field, in both Sierra Leone and the Caribbean, the tone is that of the cosmopolitan observer, at ease with tropical difference precisely to the extent that it may be set at a distance and subjected to the

FIGURE 6.7 *A*, Henry Smeathman, *The Termitaries of "Termes atrox"*: original sketch (©Royal Society Archives, London). There is a note on the original sketch, above the figure to the right: "This group and the figure not to be engraved." *B*, Henry Smeathman, *The Termitaries of "Termes atrox,"* as engraved by James Basire (*Philosophical Transactions* 71 [1781]), with the figure removed.

comparative gaze of the naturalist. The eye that observes, compares, and contrasts is itself rarely visible. In this context, it is worthy of note that the published version of one of Smeathman's sketches of "turret nests" (*Termes atrox*) omits a miniature vignette of an observer—presumably, the naturalist himself—which was part of the original drawing (fig. 6.7). Written faintly in pencil on the original sketch are the words, "This group and the figure not to be engraved." We are tempted to attribute this editorial intervention to Joseph Banks, who had assumed the presidency of the Royal Society in 1778. Whether this was actually the case is hardly the point: there was only room for one presiding genius.

If the details of Smeathman's own biography and his career as a naturalist have slipped from the view of most historians, his sketches of tropical termitaries have taken on a life of their own. In particular, his representation of the habitations of *Termes bellicosus,* combining the composed picturesque scene, the dissecting analytical eye, and the synoptic landscape view within a single frame has been a lasting presence in the field of entomology. Hence, in a recent scientific paper celebrating "two hundred years of termitology," the *Philosophical Transactions* engraving was reproduced in a double-page illustration.[54] Copies and reworkings of Smeathman's cross-sectional views may today be found in numerous entomology textbooks and museum displays around the world.[55] The very sketches that were supposed to substitute the empirical for the symbolic have themselves become iconic representations of tropical nature. Where images of tropicality are concerned, pictures often speak louder than words.

★ 7 ★

Matthew Fontaine Maury's "Sea of Fire": Hydrography, Biogeography, and Providence in the Tropics

D. GRAHAM BURNETT

Let me begin with a footnote. Sometime in the summer of 1851, Herman Melville inserted a last-minute addendum into the manuscript of his sprawling saga of man against beast—the book known variously as *The Whale, The White Whale,* and finally and forever, as *Moby-Dick.* This footnote fell in a key chapter, entitled "The Chart," where Melville takes on a (possibly the) central implausibility in his tale: How is it that Captain Ahab proposes to reencounter a single cetacean (albeit very large and strangely colored) in the vasts of the world oceans? What are the chances? This is no small matter, dramatically speaking: after all, the whole point of *Moby-Dick* is to tell the story of Ahab "chasing" the white whale. Yet this presents a serious problem: Just how, exactly, can one chase a whale?

Melville takes on the reader's skepticism in "The Chart." And his answer there casts the book's notorious protagonist in a curious new light. For in "The Chart," Melville shows us that Ahab, despite his surging, demonic passions, is a creature of incisive rationality, of methodical and empirical mind—in fact, this monomaniacal figure is, we discover, nothing less than a student of the physical geography of the sea and its denizens, a practitioner of some of the most sophisticated natural science of his day. I quote from the opening of the chapter:

> Had you followed Captain Ahab down into his cabin after the squall . . . you would have seen him go to a locker in the transom, and bringing out a large and wrinkled roll of yellowish sea charts, spread them before him on his screwed down table. Then, seating himself before it, you would have seen him intently study the various lines and shadings which there met his eye; and with slow but steady pencil trace additional courses over spaces that

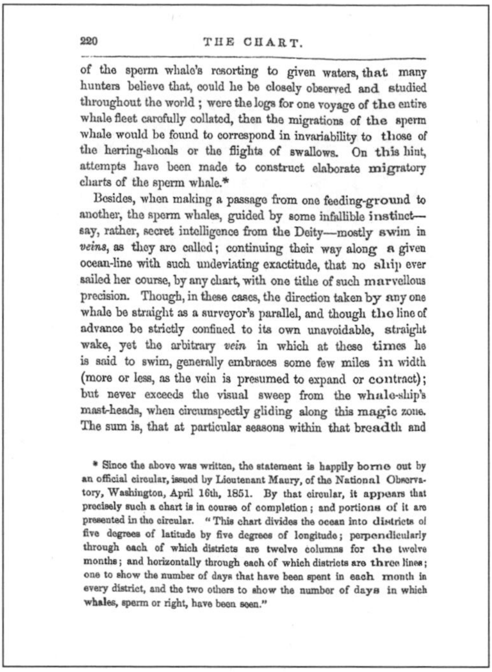

220   THE CHART.

of the sperm whale's resorting to given waters, that many
hunters believe that, could he be closely observed and studied
throughout the world; were the logs for one voyage of the entire
whale fleet carefully collated, then the migrations of the sperm
whale would be found to correspond in invariability to those of
the herring-shoals or the flights of swallows. On this hint,
attempts have been made to construct elaborate migratory
charts of the sperm whale.*

    Besides, when making a passage from one feeding-ground to
another, the sperm whales, guided by some infallible instinct—
say, rather, secret intelligence from the Deity—mostly swim in
veins, as they are called; continuing their way along a given
ocean-line with such undeviating exactitude, that no ship ever
sailed her course, by any chart, with one tithe of such marvellous
precision. Though, in these cases, the direction taken by any one
whale be straight as a surveyor's parallel, and though the line of
advance be strictly confined to its own unavoidable, straight
wake, yet the arbitrary *vein* in which at these times he
is said to swim, generally embraces some few miles in width
(more or less, as the vein is presumed to expand or contract);
but never exceeds the visual sweep from the whale-ship's
mast-heads, when circumspectly gliding along this magic zone.
The sum is, that at particular seasons within that breadth and

    * Since the above was written, the statement is happily borne out by
an official circular, issued by Lieutenant Maury, of the National Observatory, Washington, April 16th, 1851. By that circular, it appears that
precisely such a chart is in course of completion; and portions of it are
presented in the circular. "This chart divides the ocean into districts of
five degrees of latitude by five degrees of longitude; perpendicularly
through each of which districts are twelve columns for the twelve
months; and horizontally through each of which districts are three lines;
one to show the number of days that have been spent in each month in
every district, and the two others to show the number of days in which
whales, sperm or right, have been seen."

FIGURE 7.1 A page from
Moby-Dick (Rare Books
and Special Collections,
Princeton University Library,
Princeton, NJ)

before were blank. At intervals he would refer to piles of old log books
beside him, wherein were set down the seasons and places in which, on
various former voyages of various ships, sperm whales had been captured
or seen.

Melville gives us to understand that Ahab pursues the white whale by calculation, using the tools of ocean biogeography: collating the data of previous voyages and correlating the patterns with the fluxes of the ocean system. Coming to the crux of the chapter, Melville continues:

> Now, to anyone not fully acquainted with the ways of the leviathans, it
> might seem an absurdly hopeless task thus to seek out one solitary creature
> in the unhooped oceans of this planet. But not so did it seem to Ahab, who
> knew the sets of all the tides and currents; and thereby calculating the drift-
> ings of the sperm whale's food; and, also, calling to mind the regular, as-
> certained seasons for hunting him in particular latitudes; could arrive at rea-
> sonable surmises, approaching almost to certainties, concerning the
> timeliest day to be upon this or that ground in search of his prey.

Now what are we to make of this? Literary license? A liberal greasing
of a sticky spot in the plot? Some elaborate Melvillian allegory?[1] Enter the
footnote of mid-1851 (fig. 7.1): "Since the above was written, the state-
ment is happily borne out by an official circular, issued by lieutenant

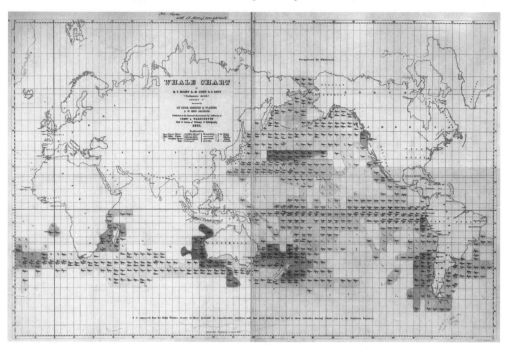

FIGURE 7.2 M. F. Maury, *Whale Chart (Preliminary Sketch),* 1851 (Library of Congress, Washington , DC)

Maury, of the National Observatory, Washington, April 16th 1851. By that circular it appears that precisely such a chart is in course of completion; and portions of it are presented in the circular." Here, as elsewhere in *Moby-Dick,* fact and fiction perform an engrossing pas de deux. For this circular, and the maps it announced, were very real indeed. Figure 7.2 and plate 8 (a detail), reproduced from the collection of the Library of Congress, show the "Preliminary" Maury *Whale Chart* published later that year, 1851.[2]

I will return to these charts at the conclusion of this chapter, with the aim of showing not only how they were made but also what surprising relevance they have to the study of the tropics in the nineteenth century, to the history of Victorian exploration, and to the science of whales—cetology—in the period. But to get to these questions, I propose that we follow up on the footnote: Who was this "Lieutenant Maury," made immortal in the small print of the great American novel? In the next two sections, I take up the answer, offering an introduction to the life and work of Matthew Fontaine Maury, while sketching the significance of his scientific investigations. While these sections primarily review an existing biographical and critical literature, I go on to suggest the need for a

FIGURE 7.3 *Portrait of M. F. Maury*, ca.1853, Bendann Studio (Frances Leigh Williams, *Matthew Fontaine Maury: Scientist of the Sea* [New Brunswick, NJ: Rutgers University Press, 1963])

revised reading of Maury's accomplishments in light of recent work in the history of nineteenth-century science. In the third and closing section, I take up Maury's understanding of the earth's tropical zones. What meaning did he attach to this distinctive social and geographical space? This final investigation will, I hope, allow us to return to the whale charts with new eyes.

MATTHEW FONTAINE MAURY: THE LIFE

Matthew Fontaine Maury is, as it turns out, considerably more than an historical footnote, particularly in the history of science (fig. 7.3, portrait circa 1853).[3] In fact, he was probably the single most decorated American man of learning in the nineteenth century—feted by European scientific societies and a familiar of their royal patrons from London to Paris, Brussels to St. Petersburg. For what? For his work, over three decades, in the naval sciences, in meteorology, and—most important—in the nascent discipline that Alexander von Humboldt called (in a letter to Maury praising his innovative publications) "The Physical Geography of the Sea." Maury borrowed the coinage as the title for his synthetic and best-remembered book, published in 1855, and this landmark volume went though eight U.S. editions in six years, staying in print in England though perhaps nineteen editions, stretching well into the 1880s; in the meantime the book had been translated into most major European languages.[4] *The*

FIGURE 7.4 Naval observatory, ca.1845 (Charles Lee Lewis, *Matthew Fontaine Maury: The Pathfinder of the Seas* [Annapolis, MD: U.S. Naval Academy, 1927])

Physical Geography of the Sea has won for Maury an obligatory nod in any history of oceanography, a science that not infrequently invokes him as its "founding father"—particularly in the United States.

But the Virginia-born Maury, who hailed from a fallen-on-hard-times branch of a distinguished American family, was emphatically not a university man (pace his honorary degrees from Cambridge and elsewhere). Rather, Maury first investigated the winds, stars, and seas during his teenage years as a midshipman in the U.S. Navy in the 1820s, shooting lunar distances from the deck of one of the young nation's six frigates, his naval commission taken up in flight from hardscrabble farming in the hills of Tennessee. A country boy and autodidact, Maury's more sophisticated hydrographic interests grew with (and, to a degree, resulted in) his impressive suite of naval promotions: from sailing master to lieutenant, from there to superintendent of the Navy's Depot of Charts and Instruments, and finally to his most illustrious post, as the first superintendent of the U.S. Naval Observatory in Washington D.C. (fig. 7.4).[5] On his way up through the tumultuous rank-politicking of the peacetime post-1812 navy, Maury authored a significant treatise on the theory and practice of navigation, a text that became, by 1843, the standard reference work and teaching tool for both the navy and the merchant marine in the United States.[6] In addition, he published a several articles on navigation and instrumentation that appeared in the emergent American scientific periodicals of the nineteenth century.[7]

Maury's national and international standing probably reached its peak in the middle of the 1850s, during the watershed years of the steamship era, when he attained significant trans-Atlantic stature as a naval hydrographer-diplomat: his pioneering investigations of the seafloor produced some of the very earliest bathymetrical charts based on actual ocean soundings, and this work earned him the company (and patronage) of wealthy projectors scheming to lay the Atlantic cable (who were hungry for pictures of the submarine world, into which they were pouring good money).[8] At the same time, the observatory under his direction was churning out elaborate global charts of wind, current, rain, temperature, and the like.[9] In this flush, in 1853, he called for and led an international meeting in Brussels that drew together a dozen representatives of the hydrographic departments of world's major ocean-going nations, who convened to draft a uniform system for maritime meteorology. This meeting, and Maury's subsequent network of correspondents, placed him more or less at the center of what he would call "the most extensive system of philosophical observations, physical investigation, and friendly co-operation that has ever been set on foot."[10] For all the truth in the boast, the phrasing was infelicitous, since this global system of observation had hardly been "set on foot"; rather, it had been set on keel. As a result of Maury's promotional and centripetal energies, hundreds of ships under the flags of half a dozen nations were, by the mid 1850s, collecting information about oceanic conditions—observing position, water temperature, prevailing winds, barometric pressure, and a host of other physical characteristics of the sea—and inscribing these measurements in "Abstract Logs" (standardized by Maury and disseminated through the work of the 1853 Brussels Conference; fig. 7.5). These were then sent to the U.S. Naval Observatory, where Maury and his staff collated the information into thematic charts, made available to everyone who participated in his expanding network. As he put it: "Rarely before has there been such a sublime spectacle presented to the scientific world: all nations agreeing to unite and co-operate in carrying out one system of philosophical research with regard to the sea. Though they may be enemies in all else, here they are to be friends. Every ship that navigated the high seas, with these charts and blank abstract logs on board, may henceforth be regarded as a floating observatory, a temple of science."[11] Maury could turn a rousing phrase, which helps account for the popular success of his books about the sea. And he certainly had a capable grasp of self-promotion.

There remains some uncertainty about the real degree of international cooperation Maury achieved under this scheme. Hearn suggests that two hundred thousand copies of the *Wind and Current Charts* (and twenty thousand copies of the accompanying volume *Sailing Directions*) were distributed free to correspondents between 1848 and 1861 and that during the fifty years after the 1863 conference some 27.5 million abstract logs

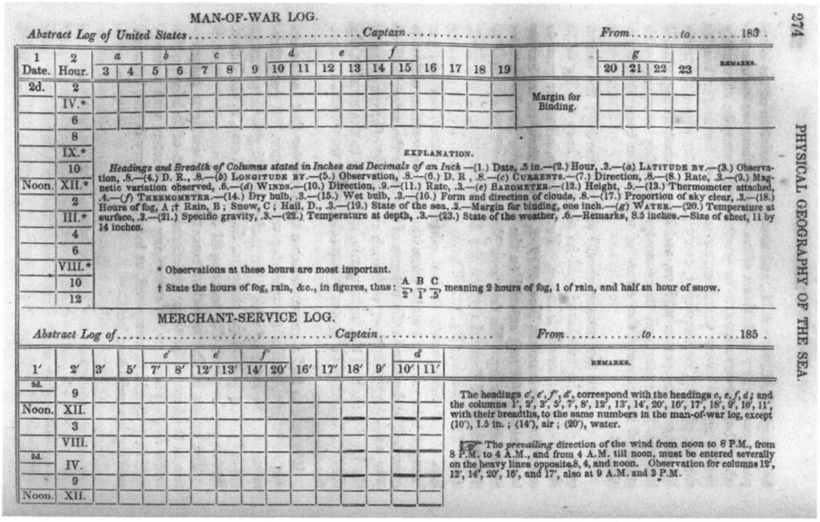

MAN-OF-WAR LOG.

Abstract Log of United States.................Captain.............. From.......to.......185.

1 Date.	2 Hour.	a		b		c		d		e		f								g			REMARKS.
		3	4	5	6	7	8	9	10	11	12	13	14	15	16	17	18	19	20	21	22	23	

EXPLANATION.

Headings and Breadth of Columns stated in Inches and Decimals of an Inch.—(1.) Date, .5 in.—(2.) Hour, .3.—(a) LATITUDE BY.—(3.) Observation, .8.—(4.) D. R., .8.—(b) LONGITUDE BY.—(5.) Observation, .8.—(6.) D. R. ,.8.—(c) CURRENTS.—(7.) Direction, .8.—(8.) Rate, .3.—(9.) Magnetic variation observed, .6.—(d) WINDS.—(10.) Direction, .9.—(11.) Rate, .3.—(e) BAROMETER.—(12.) Height, .5.—(13.) Thermometer attached, .4.—(f) THERMOMETER.—(14.) Dry bulb, .3.—(15.) Wet bulb, .3.—(16.) Form and direction of clouds, .8.—(17.) Proportion of sky clear, .3.—(18.) Hours of fog, A ;† Rain, B ; Snow, C ; Hail, D., .3.—(19.) State of the sea, .3.—Margin for binding, one inch.—(g) WATER.—(20.) Temperature at surface, .3.—(21.) Specific gravity, .3.—(22.) Temperature at depth, .3.—(23.) State of the weather, .6.—Remarks, 8.5 inches.—Size of sheet, 11 by 14 inches.

* Observations at these hours are most important.

† State the hours of fog, rain, &c., in figures, thus: $\frac{A \; B \; C}{2 \; 1 \; .5}$ meaning 2 hours of fog, 1 of rain, and half an hour of snow.

MERCHANT-SERVICE LOG.

Abstract Log of.................Captain.............. From.............to.............185 .

1′	2′	3′	5′	7′	8′	12′	13′	14′	20′	16′	17′	18′	9′	10′	11′	REMARKS.

The headings c′, e′, f′, d′, correspond with the headings c, e, f, d ; and the columns 1′, 2′, 3′, 5′, 7′, 8′, 12′, 13′, 14′, 20′, 16′, 17′, 18′, 9′, 10′, 11′, with their breadths, to the same numbers in the man-of-war log, except (10′), 1.5 in. ; (14′), air ; (20′), water.

☞ The *prevailing* direction of the wind from noon to 8 P.M., from 8 P.M. to 4 A.M., and from 4 A.M. till noon, must be entered opposite on the heavy lines opposite,8, 4, and noon. Observation for columns 12′, 13′, 14′, 20′, 16′, and 17′, also at 9 A.M. and 3 P.M.

FIGURE 7.5 Abstract log (M. F. Maury, *Physical Geography of the Sea,* 2d ed. [1855])

had been turned in by seamen, with the Germans and the British leading the way (with 10.5 and 7 million, respectively).[12] On a preliminary investigation, however, these numbers do not inspire great confidence. My sense is that the vast majority of Maury's "floating observatories" were, in fact, U.S. ships.[13]

Nevertheless, the cartographic results of Maury's oceanic observational network cannot be overlooked: more than one hundred elaborate foliosized thematic charts, published over more than a decade, collating millions of individual observations, printed in more than one hundred thousand exemplars, and used by ships under numerous flags. Perhaps the best proof of the importance of these researches lies in Maury's ties to the deep-pocketed world of international insurance finance, the maritime underwriters from Lloyd's of London to the Tontine Coffeehouse on Wall Street. These financiers saw in Maury's thematic charts the promise of safer shipping (as well as more rational risk assessment); they would remain his staunchest allies (and patrons) even as his career unraveled in the 1860s.[14]

Before turning to look at this charting work in greater detail, let me conclude this brisk biographical sketch by tracing Maury from his apogee down the unhappy slope to his demise. Even as his navigational texts and charts were winning him international kudos in the later 1850s, rivals and detractors at home were conspiring to undermine Maury's position as the celebrated "pathfinder of the seas." Nor was this opposition without its justifications. For many in the emerging university scientific establishment in the United States, Maury had become an impediment at best and, at worst, a national embarrassment. For there, at the helm of the "national" observatory (one of a mere handful of government scientific positions) stood a man best characterized—say, from the vantage point of a chair in natural philosophy at the University of Pennsylvania or the College of New Jersey (later Princeton University)—as an "Old Salt": an energetic hand with a sextant, yes; a man who looked good in a uniform, perhaps; and someone who could appreciate a carefully engraved nautical chart, fine; but not a proper "astronomer," not an astronomer capable of making original contributions to, say, celestial mechanics, or even capable of producing a star atlas to stand beside those of Greenwich or the Paris Observatory.[15]

This galled men like Joseph Henry (the head of the nascent Smithsonian Institution) and his ilk, the nucleus of the newly forming community of professional scientists in the United States: Alexander Dallas Bache (director of the U.S. Coast Survey), for instance, and Benjamin Pierce (professor of mathematics and director of the Harvard Observatory).[16] On top of the ire of these men of science, Maury's rapid promotions (at least in the eyes of rivals; he himself thought them rather sluggish), and his active career as a naval reformer–cum-gadfly (he was continuously writing thinly veiled anonymous attacks on naval bureaucracy and deadwood officer corps) had succeeded in earning him a slate of serious enemies in the navy as well, each of whom looked forward to seeing him descended. Not least of these was the "stormy petrel" himself, Charles Wilkes, Maury's predecessor at the Depot of Charts and Instruments and, subsequently, commander of the most elaborate scientific enterprise sponsored by the U.S. government in its first seventy-five years, the U.S. Exploring Expedition, or U.S. Ex. Ex.[17]

In the end, the opposition of Maury's scientific and naval detractors played no role in his spectacular fall from grace. It was the Civil War that would be his undoing. On 19 April 1861, shortly after the bombardment of Fort Sumpter, Maury learned that the State of Virginia had seceded from the Union. That night, in anguish, he gathered his belongings from the observatory, and prepared to cross the Potomac to return to his native state. The next morning he drafted his letter of resignation from the navy, a letter that his daughter would call, colorfully, but not falsely, "the death warrant to his scientific life—the cup of Hemlock that would paralyze and kill . . . his pursuit after the knowledge of nature and nature's laws."[18]

His service to the Confederacy over the duration of the war consisted of quixotic and ultimately futile efforts to deflect the naval dominance of the Federal forces. Maury experimented with mines and torpedoes and ran the blockade to seek ships and munitions in England but, ultimately, watched from the sidelines as his state and his scientific stature slid into ruins. A wanted man, he stayed away from the United States after the war, selling torpedo-building miniseminars to foreign navies, and toadying to Emperor Maximilian in Mexico as part of an abortive effort to found a "New Virginia" colony of southern irredentists in the Cordova Valley. Surviving Maximilian's execution in a popular coup (he had the good fortune to be in England at the time), Maury lived by his pen, supporting his family by authoring a shelf of popular school geographies, while living in exile (and much diminished circumstances) in London. Though he returned to Virginia at the end of his life, his scientific career was, by 1873 (the year of his death), a distant memory.[19] With this sense of the arc of Maury's life in mind, let us turn to a more detailed examination of his scientific career at its height, as well as recent assessments of the place of that career in the history of science.

MAURY'S WORK AND ITS SIGNIFICANCE

I focus here on Maury's charting and hydrographic activities at the Naval Observatory in the crucial decade of the 1850s. Working my way through these materials I have at times been struck with the sense that the whole of Maury's cartographic opus, for all its originality and scope, could almost be deduced from a pair of biographical facts: first, his rivalry with Charles Wilkes, the man who conducted the glamorous (if controversial) U.S. Exploring Expedition, 1838–42, and second, a personal disaster—the crippling stagecoach accident (on 17 October 1839) that cost Maury the normal use of his right leg for the rest of his life and, in doing so, ensured that he would never set sail on a ship in the navy again. Added together, these two merely human factors sum to a considerable problem: How could the proud and ambitious Maury rival his adversary Wilkes's global multiship scientific exploring expedition without being able to sail? How to rival Wilkes and the U.S. Ex. Ex. without ever leaving a desk job in Washington?

It is a testimony to Maury's distinctive tenacity that he actually solved this problem. The answer lay in maps, like those represented in figures 7.6 and 7.7. On taking up his position at the Depot for Charts and Instruments in 1842, just a few months after the return of the U.S. Ex. Ex., Maury became aware of the potential value of the depot's stash of old logbooks from both naval and merchant marine vessels. These logs contained the day-by-day geographical positions and wind observations for hundreds of thousands of miles of ocean voyaging. Maury set his staff of midshipmen to the laborious task of collating the information in these

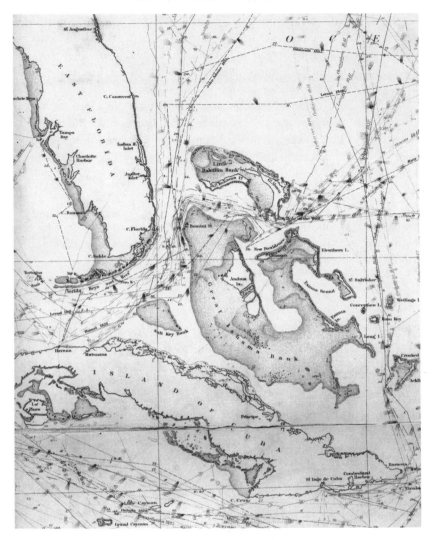

FIGURE 7.6 Detail from a track chart: Cuba and the Bahamas (Princeton University Library, Princeton, NJ)

logbooks, and the work took shape in charts like the one shown here, a detail from a much larger sheet depicting the central Atlantic.[20] It is difficult to communicate the magnitude of this operation. Over the subsequent decade, Maury's track charts, as they came to be called, spanned the entire globe in breathtaking detail—about forty large charts in all, recording the daily positions and wind observations of hundreds and hundreds of ships according to a seasonal color code; the Atlantic Ocean alone

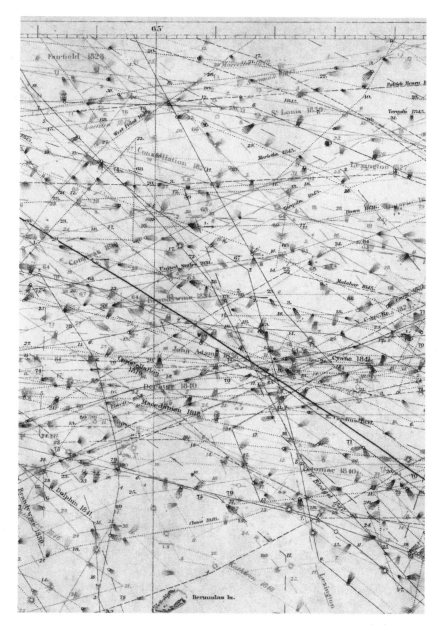

FIGURE 7.7 Detail from a track chart: Atlantic, north of Bermuda (Princeton University Library, Princeton, NJ)

FIGURE 7.8 Example of the pilot chart (Princeton University Library, Princeton, NJ)

covered eight sheets that, if laid out, covered an area six feet high by eight feet wide. Paging through the folders of these charts offers a remarkable view of the early nineteenth-century maritime world.

This enterprise bore fruit rapidly: on completing the Atlantic coast track charts, Maury noticed that American captains running the popular New York to Rio route had a tendency to swing very wide toward Africa before picking up the Southeast trade winds and zigzagging back to the west. Rumor had it that dangerous currents and unfavorable winds hugged the eastern shore of the South American mainland, making the more direct route a perilous enterprise. Maury thought his track charts told otherwise. In 1848, Captain Jackson of the bark *W. H. D. C. Wright,* using Maury's directions, made the Rio passage in thirty-eight days from Virginia, shaving more than two weeks off the usual run. Then he repeated the feat on the return. Newspapers picked up the story, and Maury's charts became a maritime sensation.[21] Similar successes followed, particularly on the New York to San Francisco passage, which Maury claimed to have cut by almost 25 percent (from about 190 days)—no small affair in the age of California Gold rush.[22]

Maury next took on the task of codifying the wind observations included on the track charts: erasing the tracks and depicting only the winds—and in doing so abstracting the track-chart data into a kind of global wind computer. The result was what he called pilot charts (fig. 7.8). These again covered the world in extraordinary detail (about thirty charts in all, but in multiple editions, leading to probably double that number of sheets): in some regions these depict the oceans on a scale of a single degree to the inch.[23] One of the most interesting details of this series is the volvelle that Maury included on every chart as a computational aide to captains planning routes.[24] Lest this work seem arcane, it is worth recalling that Maury undertook these synthetic wind charts in the heady days of the final showdown between the power of sail and steam. The 1850s are remembered by maritime historians as the golden age of the clipper ships, those fastest of the fast sails, and the period saw numerous races staged to defend the honor of wind power on the high seas.[25] Maury was keenly aware of this rivalry and believed that his charts would save the age of canvas, writing in the first edition of the *Physical Geography of the Sea*: "The modern clipper ship, the noblest work that has ever come from the hands of man, has been sent, guided by the lights of science, to contend with the elements, to outstrip steam, to astonish the world."[26]

And Maury continued this cartographic project—squeezing the wind data—in two more documents that further abstracted and generalized atmospheric patterns: he called these the trade wind charts (or ser. B; figs. 7.9 and 7.10). Without delving into the minutiae of the B series, it is quite clear that Maury's interests in the winds and currents of the sea were growing more explicitly synthetic and, we might say, "theoretical" in the mid-1850s. Having now conducted a Wilkes-rivaling global exploring expedition of his own (one of unprecedented extent), having sailed the globe by proxy and pen for almost a decade, Matthew Fontaine Maury wanted to write up his results.[27] As he put it: "I find that the tracks of vessels are full of meaning."[28] And he took it as his mission to reveal and explain that meaning to a reading public.

At first this project took the form of textual addenda to the charts, a kind of longish key or legend to accompany them. These appeared as a separately published booklet in 1850, under the title *A Notice to Mariners*, but

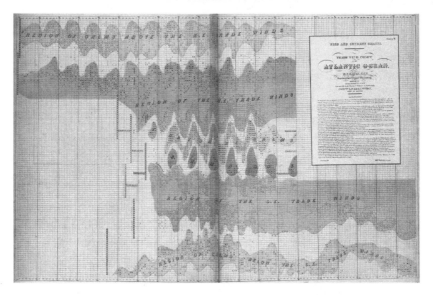

FIGURE 7.9 Detail from *Trade Wind Chart of the Atlantic Ocean,* ser. B (New York Public Library)

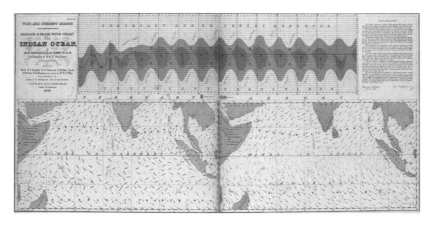

FIGURE 7.10 Detail from *Monsoon and Trade Wind Chart of the Indian Ocean,* ser. B (New York Public Library)

rapidly expanded to a 315-page book, *Explanations and Sailing Directions to Accompany the Wind and Current Charts* published in 1851, and this ballooned in subsequent editions, until by 1858 it was a two-volume, 1,300-page doorstop.

It was in these volumes that Maury tried not only to describe but also to explain the systems of global ocean dynamics that his charts had begun to show in spatialized and quantified form: the flows of currents, the seasonality of winds, the fluxes of drifting tides. And it was from these books—increasingly unwieldy accretions of suggested sailing routes, physical-geographical ruminations, and hypotheses, all larded liberally with long citations from admiring correspondents—that Maury culled the material for his landmark text, the book I have already mentioned, and the book by which he remains best known: *The Physical Geography of the Sea*. This text is, when looked at closely, little more than a pruning of the best bits from the swollen *Sailing Directions*.

There are several things worth noting about this chain of textual composition and borrowing, since it is really quite remarkable that the volume widely considered foundational in the science of oceanography began life as a cut-and-paste redaction from an oversized naval almanac, itself a hypertrophied cartographic key. Part of what makes this so striking is that Maury seems to have rush-composed the shorter work that would secure his historical reputation out of anxiety that someone else would plagiarize his magnum opus to produce a profitable popular book on the seas; Maury wanted to beat them to the punch.[29] Historians of science have grown increasingly attentive in the past few years to the place of print culture in knowledge formation, and Maury's *Physical Geography of the Sea* offers a rich case study for an examination of these issues, linking as it does, print piracy, popular science, and discipline formation.[30] That his foray into science popularization became his most significant legacy, and thereby helped to define a field of scientific inquiry, dramatizes the danger of treating Victorian popular science as merely attenuated seepage from a well-defined domain of properly scientific activity; that the very success of the book eventually helped draw him away from scientific institutions, and into the life of a writer who made his living by the pen, reflects the complex conditions that birthed professional science and professional writing as twin offspring of the mid-nineteenth century, joined (perhaps) at the hip.[31]

Maury himself sought explicitly to marry the knowledge of nature and the knowledge of the masses. To understand how, we need to return to the program of his increasingly ambitious charts: What happened when he began to try to "make sense" of what his elaborate cartographic collations revealed? What was the "meaning" that he had (as he put it) "threshed" from the logbooks?

In brief, in those lines on the chart, in the wakes of the navy cutters, Maury found—God. Maury's "physical geography of the sea" fit firmly within the established tradition of natural theology, the project that exhorted the suitably devout natural philosopher to "look through nature up to nature's God."[32] For Maury—who worshiped regularly at St. John's Episcopal Church in Washington, D.C., but who embraced a Campbellite nonconformist brother and penned a fluttering, impassioned private devotion for sustenance in adversity—teleology and biblical apologetics retained a central place in the proper investigation of nature.[33] As he put it in *The Physical Geography of the Sea:* "The theory upon which this work is conducted is that the earth was made for man," and it followed from this that nature manifested divine intelligence in all its elements, from the stem of a snowdrop flower to the meridian transit of a bright star.[34]

Maury powerfully extended this familiar interpretive framework to the vast expanses of the open ocean.[35] He attacked this formless, empty, recalcitrant space, with the intention of showing that it was a precisely calibrated and complex machine, infinitely more intricate and better-regulated than the finest chronometer, with its own compensators, mainspring, cogs, and jewels.[36] And he exemplified this brilliantly, narrating with considerable poetry the fragile-yet-powerful, steady-yet-changing dynamics of the sea, the continuous circulations caused by variations in temperature, density, and chemical combination, the providential systems that supported the great "chain of being" dwelling in the depths. As he wrote toward the conclusion: "Now we do know that . . . [the sea's] adaptations are suited to all the wants of every one of its inhabitants—to the wants of the coral insect as well as to those of the whale."[37]

It is important to note that just because Maury made sense of nature according to the framework of teleological natural history (old-fashioned by 1858, even quaint), this in no way means that he was blind to quite sophisticated physical, chemical, and biological connections in the natural world—the sort of balances and webs we now think of as "ecological." In fact, if anything, Maury was predisposed to seek out such delicate and improbable global systems precisely because their very delicacy and improbability bespoke the glory of the "Architect of Creation." It might be worth underlining this point, since it has real relevance to any broader effort to assess Maury's place in the history of science—which turns out to be quite a contested matter. A brief detour into the historiography may be of use here.

In the early 1960s, several distinguished historians of science and geography took a look at Maury and reappraised his significance, considerably downgrading his stock in the history of the ocean sciences.[38] Schoolbooks might call him the founder of oceanography, they noted, but it was not at all clear that he merited the title: he was a popularizer at best, and an inferior student of the sciences of earth, sky, and sea. There is clearly some truth to this cold-eyed review of Maury's contributions to science in general and American science in particular, and it served as a necessary corrective to a line of Maury apologists and enthusiasts who tended to overstate his achievements, decking him out as a kind of Confederate New-ton.[39] But at the same time this reassessment was rooted in a style of history of science particularly concerned to establish and trace the genealogy of the right answers to the hard questions about the natural order. Maury, to be sure, got most of the answers wrong: he thought magnetism was somehow involved in the circulation of the atmosphere, he thought that air streams "crossed" in the doldrums, and he thought winds did not really cause ocean currents. Moreover, to historians of science of a certain generation, Maury's extravagant natural theology went a long way toward demonstrating that he was not a true scientific innovator, that he was nothing like, for instance, everyone's favorite example of a mid-nineteenth-century scientific hero, Charles Darwin (who was Maury's almost exact contemporary—in fact, they quite literally "passed like ships in the night" off the Falklands in March of 1834).

But the past forty years of scholarship in the history of science have shifted our approach in a number of ways, enabling us to see Maury afresh, without resort to either Dixie hagiography or wheat-from-chaff threshing. Work like that of Susan Faye Cannon, Nicolaas Rupke, and Michael Dettelbach, for instance, has given us a better sense of the importance of Alexander von Humboldt in the first half of the nineteenth century, and there is certainly much to be said about Maury as an American Humboldtian (see chap. 1).[40] Equally important is a revised view of the durability, complexity, and significance of natural theological thought in the history of the sciences of matter, motion, and life.[41] The old story (crudely) was that the sustained study of the laws of nature—real science—emerged out of the progressive extirpation of "backward" and "blinkered" god talk: the less God, the more science. Such accounts—of conflict, maturation, and the overcoming of superstition—have now seen considerable reworking across a wide range of periods, places, and enterprises. In fact, looking more closely at certain contexts, it sometimes seems that, in profound ways, the emergence of something like science may well be entirely contingent on such god talk: facilitated by it, not hampered. How? Take Maury as an example: by the old view, Maury fails as a "real" proto-oceanographer to the degree that he indulges in natural theological reveries. But couldn't we argue precisely the reverse? By situating God at the helm of a vast sea-machine (precisely calibrated, meticulously regulated), Maury folded the oceanic environment into the realm of order and law. And this is no small thing, particularly in light of the deep Western tradition of depicting the ocean as the very opposite of order and reason, as a zone of darkness and chaos, a space behind God's back.[42] Taking this into account, we might say that only once the sea has been posited as a realm of divine beneficence, as a manifestation of God's rational plan for the cosmos, does it become possible to investigate it "scientifically," with some expectation of revealing its immanent logic.

In this view Maury's natural theological vision of the sea is not a blind ally in the history of oceanography but a legitimate, significant, and even, potentially, a constitutive episode in the history of the sciences of the sea. We can perhaps go even further, since Maury not only extended a metrical, collative natural theological framework to the dark spaces of the sea, he also extended a system of enlightening scientific observation to sailors, a notoriously benighted population.[43] Seen this way, it's possible to read Maury's *Physical Geography of the Sea* not as the failed theoretical overreaching of some sort of scientist manqué (very much the tone of several of the assessments from the 1960s) but, rather, as an extremely successful program for the moral reformation of the world's sailors through science. These tars and their officers would learn that the sea was a godly realm, and they would learn this through participating, by means of the careful use of instruments on board their ships, in nothing less than the collaborative revelation of divine order. Maury's network would transform them, turning them from hardened men of the bilge into what he called "contemplative" and "right-minded mariners."[44] If I am right about this, then *The Physical Geography of the Sea* has been significantly misunderstood, since by my argument it must actually be read as an extension of Maury's life-long work of naval reform. Improbable? Wrote one captain from the coast of Chile in 1855: "I feel that, aside from any pecuniary profit to myself from your labors, you have done me good as a man. You have taught me to look above, around, and beneath me, and recognize God's hand in every element by which I am surrounded."[45]

THE MEANINGS OF THE TROPICS

Let me turn then, in conclusion, to the questions that link *The Physical Geography of the Sea* to the theme of this volume and that will draw us back, in an ocean gyre, to *Moby-Dick*. What did Maury have to say about the tropics, and what were the implications for the history of exploration and the science of whales in the nineteenth century?

Maury was no stranger to the tropics. In his circumnavigation as a young midshipman in the U.S.S. *Vincennes,* 1827–30 — traveling with among others, Herman Melville's cousin, Thomas — Maury ran down the Pacific line, with stops in Manila, Java, Sumatra, the Marianas, Macau, Hawaii, and perhaps most significantly, the Marquesas Islands, where he had a chance to inquire after his then-deceased older brother, John, who had lived for more than a year as a beachcomber on Nuku Hiva Island, the same island where Herman Melville, a few years later, would jump his whaling ship and live out the tropical romance that would lead to the novels *Typee* and *Omoo*. Maury thus lived the first American age of tropical adventurism, but what did he think about this region, what place did it have in his vision of the oceanic-atmospheric system? An important place, it turns out, and one that nicely illustrates the theme I have addressed in the previous section: Maury's maritime natural theology.

Take, for instance, this quote from *The Physical Geography of the Sea*:

> Modern ingenuity has suggested a beautiful mode of warming houses in winter. It is done by means of hot water. The furnace and the caldron are sometimes placed at a distance from the apartments to be warmed. It is so at the Observatory. In this case, pipes are used to conduct the heated water from the caldron under the superintendent's dwelling over into one of the basement rooms of the Observatory, a distance of one hundred feet. These pipes are then flared out so as to present a large cooling surface; after which they are united into one again, through which the water, being now cooled, returns of its own accord to the caldron. . . . Now, to compare small things with great, we have, in the warm waters which are confined in the Gulf of Mexico, just such a heating apparatus for Great Britain, the North Atlantic, and Western Europe. The furnace is the torrid zone; the Mexican Gulf and Caribbean are the caldrons; the Gulf Stream the connecting pipe.[46]

Maury elaborates this analogy at considerable length, but the thrust is clear and is reiterated throughout the work: the tropics serve as the furnace and caldron of the providential atmospheric-terraqueous earth engine.[47]

But the deeper meaning of this tropical cauldron lay, strangely, far to the north. To understand how, we must return to the footnote with which we began, to Maury's whale charts (fig. 7.2 and plate 8, the version of 1851; and fig. 7.11, the version of 1853). The whale charts—series F, the final set of documents in Maury's extensive cartographic enterprise—should now make more sense.[48] They were a product of the very same process—the collation of logbooks—that produced the pilot charts. Figure 7.12 is a graphic depiction of the process of logbook collation, and there is a full world map of this form, in four sheets, which is the final fruit of the F series. In this case, the logs hailed from a selection of the more than seven hundred U.S. whalers afloat around the globe, plying their trade.[49] And here, too, as in the study of the track charts, there were material advantages to be gained—Maury was quick to point out that the whaling industry not only employed thousands of American seamen but also drew up from the deep far greater value, he asserted, than all the gold of California. Better knowledge of whaling grounds was worth a fortune.

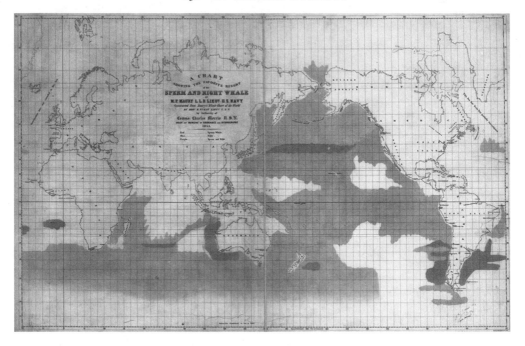

FIGURE 7.11 M. F. Maury, *A Chart Showing the Favorite Resort of the Sperm and Right Whale,* 1853 (Library of Congress, Washington, DC)

But as in the case of the track charts, there was more than merely practical knowledge to be gained from the threshing of the logs: there was hidden meaning, providential inscriptions written by the Creator in the workings of the sea system, and only legible through the painstaking cartographic collation that was the work of the observatory. Maury wrote: "Log-books containing the records of different ships for hundreds of thousands of days were examined, and the observations in them coordinated for this chart. And this chart . . . led to the discovery that the tropical regions of the ocean are to the right whale as a sea of fire, through which he can not pass, and into which he never enters."[50] So the tropical cauldron was apparently impassible to the right whale. Who cared? Well, this was, if true, a startling and significant discovery: after all, the pious whaling captain and man of science, William Scoresby Jr., had already shown (based on recovered harpoons) that whales struck in the North Atlantic were sometimes later taken in the North Pacific.[51] And this meant a remarkable chain of deductions could be forged: if right whales could appear on both sides of the Americas, but they could not cross Maury's equatorial "sea of fire," then, as Maury wrote, "we are entitled to infer that there is, at times at least, an open water communication between

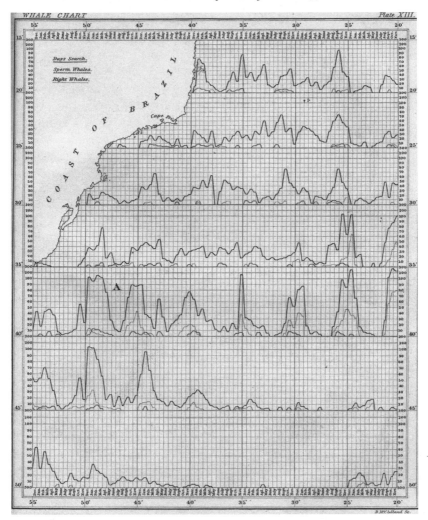

FIGURE 7.12 Whale chart, off the coast of Brazil, ser. F (Princeton University Library, Princeton, NJ)

these Straits [the Bering Strait] and bay [Baffin Bay]—in other words, that there *is* a north west passage."[52]

Here was Maury's providential geography in action, the apotheosis of his natural theological oceanic cartography. While John Franklin's bones whitened on the tundra, while his would-be savior, Robert McClure, made his ship fast to an iceberg in Mercy Bay and crossed his fingers, Matthew Fontaine Maury could claim already to have "discovered" the route they sought, discovered it by calculation, by reading between the lines of the godly sea system that he had carefully worked to plot and manifest and that he hoped would save souls in more ways than one.[53]

For Franklin, of course, it was too late, and—where souls are concerned—it is a testimony to Melville's impious genius for irony that his lost Ahab went spiritually astray hunched feverishly over a Maury-esque map of the physical and biological geography of the seas. But perhaps Maury, in the end, gets the last laugh: the logbook of Melville's own whaling voyage aboard the *Acushnet* disappeared more than a century ago, but not before it was abstracted for Mathew Fontaine Maury's whale charts. As a result, a precise record of the voyage survives, and it is now students of *Moby-Dick* who can unroll Maury's track charts and pursue the wanderings of Melville himself "in the unhooped oceans of this planet."[54]

SITES

Envisioning the Tropics: Joseph Hooker in India and the Himalayas, 1848–1850

DAVID ARNOLD

By the mid-nineteenth century, the idea of "the tropics" was firmly entrenched in the popular imagination as well as in the scientific literature of Europe and North America. One striking illustration of this was *The Heart of the Andes,* the vast painting by the American landscape artist Frederic Church, first exhibited to public acclaim in New York in 1859 (plate 3).[1] It captures pictorially many aspects of the grand vision of tropical South America previously enunciated in print by the great naturalist Alexander von Humboldt through his travel narrative and studies in plant geography. Church's painting powerfully conveys a sense of tropical vegetation as luxuriant, dense, and fecund. It suggests the supremacy of nature in the tropics, with the mountains, waterfall, and forests dwarfing the human presence (most evident in two figures by a wayside cross at bottom left of the picture and, beyond them, a small lakeside settlement glistening in the sun). It further hints at a dynamic interrelationship between the tropical vegetation in the foreground, the vast open plains in the middle distance, and, beyond, mist-shrouded hills rising into majestic snow-covered peaks. The huge scale, romantic grandeur, scenic complexity, and narrative progression in Church's painting reflect what many naturalists as well as landscape artists (and their respective audiences) had come by the 1850s to understand and value as constituting the essence of the tropics. When Charles Darwin or Joseph Dalton Hooker (the subject of this chapter) thought of the tropics in a visual sense, this was the kind of grand prospect they had in mind.

Moving from pictures to prose, as representations of tropicality commonly did in the late eighteenth and nineteenth centuries, in 1865, six years after *The Heart of the Andes* was first exhibited to the American public, the *Quarterly Review* in London published a review of five recent works titled "Natural History of the Tropics." The books chosen for review effectively spanned the tropics—from Alfred Russel Wallace and Henry Walter Bates on the Amazon, to Joseph Hooker on Bengal, Sikkim, and the Himalayas, and the Reverend W. Ellis on Madagascar, along with *The Tropical World* by Dr. G. Hartwig, a "popular scientific account of the animal and vegetable kingdoms." Predictably, the anonymous reviewer began his remarks by quoting the supreme authority, Humboldt, and gave pride of place to the scientific travelers in tropical South America. It was with seeming reluctance that, several pages into the review, he left the New World in order to "cast a glance, in company with one of the greatest botanists of the day," at "the tropical features of the Sikkim Himalayas," a region that was "not strictly speaking within the Tropics," but where, at the base of the mountains, the vegetation was "of a tropical character."[2] Only in his closing paragraph did the reviewer see fit to mention Dr. Hartwig. In the company of such famed and far-traveled naturalists, the doctor's "popular" book appeared to merit scant consideration. Yet in its almost mesmeric repetition of the word "tropical" (to describe everything from hurricanes to spiders), in its numerous woodcut illustrations (many extracted from earlier works of travel and natural history), in its quotations from Humboldt, Darwin, and Hooker (as well as such other authorities as David Livingstone on Africa and James Emerson Tennent on Ceylon), even in its very title (ninety years before Pierre Gourou), *The Tropical World* arguably did most among the five books under review to demonstrate how rapidly the idea of the tropics had become domesticated in the Euro-American imagination and rendered materially and discursively accessible to the West.[3] But what contribution did Joseph Hooker, lauded by the reviewer as one of the greatest botanists of his day, make to the scientific and wider public understanding of the tropics and for a region that was "not strictly" within the tropics and yet paradoxically possessed vegetation "of a tropical character"?

TROPICAL SOUTH ASIA

India, which Hooker visited for the first time in 1848, was in the tropics but not necessarily of the tropics. The maritime voyages of Louis-Antoine de Bougainville, James Cook, William Bligh, and others that had helped to first fix the idea of the tropics in European art, natural history, and ethnography had all passed India by.[4] Darwin had not touched its shores nor had Humboldt climbed its highest mountains. The language of tropicality, and both the visual and the scientific representation of that con-

FIGURE 8.1 *The Challees Satoon in the Fort at Allahabad on the River Jumna,* from Thomas Daniell and William Daniell, *Oriental Scenery: Twenty-four Views in Hindoostan* (London, 1795), plate 6 (by permission of the British Library)

tradictory idea, torn between paradise and pestilence, had until about the 1830s been far more commonly applied to the Pacific Islands, the Caribbean, and equatorial America than to the Indian subcontinent.[5] East India Company surgeons were familiar with many of the diseases that would in time be designated "tropical," but in the early nineteenth century they were still inclined to consider them the consequences of "warm climates" or locally specific environmental and miasmatic causes.

Surgeons aside, India had its professional artists, usually itinerant, embarked on the north Indian equivalent of Europe's Grand Tour, but the landscape they depicted was often strikingly different from overtly "tropical" visions of the South Seas and elsewhere. Perhaps reflecting consumer taste as much as artistic sensibilities, their scenes tended to be Orientalist and picturesque rather than romantic and sublime. In the paintings of William Hodges, who traveled in India between 1778 and 1784, after having been on Cook's second Pacific voyage, or the familiar aquatints of Thomas and William Daniell, (the uncle and nephew team of illustrators who visited India from 1786 to 1794), these Orientalist landscapes focused on the monuments that littered the Indian plains, on ancient temples and mosques, forts, and tombs, set against a dull, khaki-colored landscape, seemingly parched by drought or ravaged by recent warfare (fig. 8.1, and chap. 2).

FIGURE 8.2 *View Taken between Natan and Taka Ca Munda, Sirinagur Mountains,* from Thomas Daniell and William Daniell, *Twenty-four Landscapes* (London, 1807), plate 21 (by permission of the British Library)

There were occasional Calcutta streets and bustling river scenes, but almost always in these Oriental scenes, nature was relegated to the background or used selectively to highlight the grandiose schemes and ruined works of man.[6] Seldom was nature foregrounded or given a pictorial identity all its own (but see fig. 8.2). When nature did hold sway, as in the elegant but utilitarian botanical illustrations to such works as William Roxburgh's *Plants of the Coromandel Coast,* it was in the form of individual plant specimens, with little attempt to capture, whether in pictures or in prose, the physical environment and scenic context in which such exotic flora occurred.[7]

By the 1840s, however, India's place in the tropics had become more assured. Medical texts dwelt with increasing frequency on the perils of a tropical climate and the importance of observing tropical hygiene.[8] Naturalists, often inspired by Humboldt, began to distinguish between tropical and temperate plants and zones of vegetation in South Asia.[9] And even general articles about European life in India or about canal and railway construction made casual reference to the tropics as an environmental hazard to be endured or overcome.[10] The stage seemed set for a more scientific and systematic exploration of India's tropicality, and for this task, Joseph Dalton Hooker, already (at thirty years old) a traveled and professionally esteemed naturalist, seemed particularly well-suited.

The son of Sir William Hooker, the director of Kew Gardens, and friend and confidante of Charles Darwin, Joseph Hooker arrived in India in January 1848 to begin a scientific journey that was to last until his eventual return to England in March 1851. Having previously served as onboard naturalist with Sir James Ross's expedition to the Antarctic, Hooker was keen, with Humboldt's blessing, to undertake a similar exploration, this time of tropical flora. Selecting the relatively unexplored eastern Himalayas, rather than the Andes whose natural phenomena Humboldt had done so much to bring to the attention of the international scientific community, Hooker traveled overland from Calcutta, through Bengal and Bihar, and into the Himalayas. From his base at Darjeeling, he explored, sketched, mapped, and botanized Sikkim, as far north as the border with Tibet and around the slopes of Kangchenjunga, then reckoned to be the highest mountain in the world. Refused permission to enter in Nepal in 1850, he embarked on a botanical expedition to the Khasi hills of Assam instead.[11] Hooker's detailed account of his travels (in letters and diaries, as well as the published version of his Himalayan journals, which appeared in two volumes in 1854 and was republished in a single-volume edition in 1891) provides rich data for an early Victorian naturalist's multilayered understanding of the tropics, for their aesthetic form as much as their scientific significance.

Mountain vegetation and topography came to dominate Hooker's published account of his travels, but it is necessary to begin where he began, with the sea. We are now so accustomed to arriving in new places by air that we have largely lost sight of the importance of the sea and coastal views to the travelers of an earlier age, not only in framing their physical approach to the land but also in informing their intellectual and emotional approach as well. This was particularly significant for the globe-girdling naturalists of the late eighteenth and early nineteenth centuries. New continents and unknown islands, which announced themselves first by sea, created impressions that were not then easily erased. Accounts of the tropics commonly began with descriptions of arriving (often, with heightened effect, at dawn) at some small harbor against the dramatic backdrop of dense tropical forests and to the accompaniment of the strange sounds of tropical birds and animals.[12] Interestingly, in the published version of his journey, Hooker gave almost no account of his voyage via the Mediterranean, Egypt, the Red Sea, Ceylon (Sri Lanka), and Madras, beginning his narrative instead with his arrival at Calcutta. The reason for this, though not apparent in print, is clear. His father, eager to publicize his son's botanical expedition to India, had printed extracts from Joseph Hooker's private letters home. When these appeared, an anonymous writer in the *Athenaeum,* a leading journal for the arts and sciences in

London, subjected them to a withering review, suggesting that Hooker had (to judge by his letters) nothing new to add to botanical knowledge, nor any talent as a travel writer. "There must," the reviewer declared, "be something very extraordinary in the style of the writer—either the humour must be great or the pictorial power unrivalled— . . . [to] command attention to such subjects. Dr. Hooker has no pretensions of the kind. . . . Besides," he added, "ordinary Englishmen" now knew the route to India "nearly as well as that to the Isle of Wight," and Hooker's description of Aden apart, the rest of his voyage to Calcutta was "about as interesting as a voyage to Gravesend would be." [13] Clearly, if Hooker were to make his mark as a scientific travel writer he would need to do more to envision the tropics for the entertainment and edification of his readers.

Nevertheless, despite the critics, Hooker's account of first arriving in South Asia is important to understanding how he viewed, and in turn represented, the tropics. His description of arriving at the port of Galle in Ceylon in late December 1847 is fairly typical of European travel writing at the time, though, with the added insights of a naturalist, it has an almost self-consciously pantropical character about it. "At Point de Galle we lay in a pretty little cove, surrounded by dense forests and wooded hills, the beach fringed with groves of Cocoa-nut Palms, and backed by forests of tropical trees of the greatest beauty. A more charming spot I never was in, reminding me altogether of the scenes described in Paul and Virginia." After a brief description of the Singhalese, he continued:

> Their houses are huts thatched with Palm-leaves, buried in groves of Co-coa-nuts and Areca or Betel-nut Palms, each cottage being overshadowed by the ample foliage of the Bread-fruit tree, one of the most luxuriant-looking trees of the tropics, thick and umbrageous, with dark green glossy leaves, and at all seasons laden with its noble fruit. The Plantain and Banana, too, are abundant everywhere, and the Pine-Apple springs up by the road-side, bearing excellent fruit, very little inferior to that grown in our English stoves. Flowers there are of all kinds, from the gaudiest and gayest to the most humble and delicate: butterflies, beetles, and gay birds all abound, and all one longs for is the bracing air and far more wholesome, though less at-tractive, beauties of an English country scene. These are nice places to see, but not to dwell in, as the pale yellow, and all but sickly faces of the English children too plainly tell. Mosquitoes and sand-flies are rife, and so are de-testable leeches, that get inside one's boot. Snakes, too, are said to be fre-quent, though I saw none of them.[14]

Along with Tahiti and Mauritius, Ceylon had an almost paradigmatic status as an island paradise. There is, nonetheless, even in this brief passage, abundant evidence for how a Victorian naturalist saw and described his encounter with the tropics. It is, first of all, striking how quickly, in a passage that emphasizes the sensuous qualities of the tropics, Hooker passes from the idyllic attributes of the tropical landscape (the fruit-laden trees, the brightly colored birds and butterflies) to the unpleasant and potentially perilous (the "detestable leeches," the unseen snakes, and the yellow-faced children). After even a short acquaintance (Hooker has doubtless been briefed by his traveling companions and island hosts), he is able to pronounce that, from a European perspective, the tropics are "nice places to see, but not to dwell in" (the point is further emphasized when he goes on to describe the Singhalese as "treacherous" and "untrustworthy," even though they appear "happy, cheerful and contented"). Indeed, throughout his travels in South Asia, Hooker's celebration of the scenic and floristic glories of the tropics is tempered (even more so in his private correspondence than in his published journals) by an acute fear of incapacitating disease and active dislike for many of the "native" inhabitants). This fear of tropical diseases was not a mere abstraction: several of his India acquaintances (and a much-loved uncle) died while he was in the country, and Hooker abandoned his own original plan to visit Borneo after India, doubting that his own health would survive that intensely tropical environment intact.

Second, the perception and experience of the tropics is always, in the mind of the traveler, situated relative to something else, mostly commonly Britain and temperate Europe. Here in a short passage about the tropics (admittedly intended for his family and fiancée rather than for the scientific community at large) we have "English children," "English stoves," and "English country scenes" to mark out the comparison. Third, it is noticeable that for an educated Victorian like Hooker—even if tropical regions have never been visited before in person (in fact, on his Antarctic voyage he had already passed through them, landing on the Brazilian coast and visiting some of the Atlantic islands)—they have been already encountered countless times in the imagination. Even for a scientist striving, one might assume, for objectivity, observation was informed by prior association. Typically, Hooker invokes (in writing to his fiancée, Frances Henslow, rather than to his father) the novel *Paul and Virginia* by Bernardin de Saint-Pierre. With its Rosseauian representation of nature and passion in Mauritius, this was one of the most popular works of literary tropicality (and an early favorite of Humboldt's).[15] But Hooker has also had a foretaste of the tropics in the pineapples grown in an English greenhouse: like Darwin and other naturalists he can immediately recognize the sight and feel of the tropics from familiarity with palm houses and heated conservatories at Kew Gardens and elsewhere. Just as botanic gardens, the natural history of Cook's voyages, and the paintings of William Hodges had helped a generation earlier to inspire Humboldt's tropical zeal, so Hooker had grown up with novels (like *Paul and Virginia*) and travel accounts (such as Cook's Pacific voyages and Samuel Turner's Tibetan embassy of 1783) that had fueled his imagination of tropical and exotic geographies from childhood.[16]

Much of what Hooker saw and recorded on his South Asian journey was (typically enough in the romantic imagination) built on a complex network of associations—things personally experienced and recalled or drawn second-hand from pictures and the printed word. Whether Hooker described tropical scenes or named new plant species (as he frequently did in the Himalayas) after his friends and patrons, he did so with a host of literary or personal associations in mind.[17] These associations could amount to a kind of nostalgia, especially when they evoked scenes of home and childhood; but to many early nineteenth-century writers, especially those of a scientific inclination, such associations were a means of registering difference as well as similarity, a device for making mental comparisons between two different places, sets of people, or personal experiences that resembled each other in certain respects but also held revealing differences ("the violation of the details," as Hooker once put it) that might repay further analysis.[18] In an exchange that echoes this, in 1854, shortly after Hooker's *Himalayan Journals* were published, Darwin asked his friend whether the "tropical vegetation" of Brazil (which he was familiar with from the voyage of the *Beagle*) was "as beautiful or nearly as beautiful" as that of the lower Himalayas. Hooker replied that that of Sikkim appeared "uncommonly fine" when compared to the plains of India, but that it did not match that of Brazil, the Khasi Hills, or Chittagong. He added that the more he read and traveled the more convinced he became "that our impressions are . . . the effects of association."[19]

BOTANY AND THE HIMALAYAS

After Ceylon and that first alluring taste of tropical South Asia, Hooker found the approach to Calcutta through the Sundarbans "exceedingly disappointing," with no evidence of the long anticipated "tropical luxuriance."[20] Calcutta offered little of interest to him and his dry-season journey in the early months of 1848 through Bengal and Bihar brought further discouragement. No Orientalist, he could muster little interest in the monuments and cities of the plains: he found Benares "a horrid place," full of dreadful noises, sights, and smells.[21] The landscape appeared parched and barren, and his mental associations were more often with bleak Scottish moorlands than with the grand tropical vistas Humboldt had helped to inspire. There were hardly any palms (so emblematic of the

FIGURE 8.3 *Kinchinjunga from Mr Hodgson's Bungalow, Dorjiling,* from J. D. Hooker, *Himalayan Journals* (London, 1854) (by permission of the British Library)

tropics) to be seen. At best there was some "low stunted jungle," unhealthy even in the dry season, and occasional "picturesque" scenes with trees that might grace an English park, but the "prime elements of a tropical flora" were "wholly wanting."[22] It was, he lamented, "altogether a country as unlike what I had expected to find in India as well might be."[23]

It was only when he began to move up from the low-lying malarial Tarai into the Himalayan foothills (outside the tropics formally defined) that Hooker began to find the luxuriant vegetation for which he was looking. He settled at Darjeeling, then in its early days as a British hill station and sanatorium, surrounded by densely wooded hills and valleys and with a commanding view of Kangchenjunga (fig. 8.3). This became the base for his botanical expeditions into Nepal and Sikkim, and over the following two years of observation and travel, he began to find the kinds of plant life and vistas that his reading and his imagination had long led him to expect. He stayed for much of his time in Darjeeling with Brian Houghton Hodgson, a leading Orientalist (whose scholarship had helped unlock Buddhism for the West) and also a self-taught zoologist and ethnographer of the Himalayas. Hodgson and other local informants enabled Hooker to graft onto his naturalist narrative and detailed botanical and topographical descriptions an appearance of ethnographic authority and a picturesque appreciation of the Buddhist monasteries and shrines that were such a prominent feature of the Himalayan landscape. Here, too, was further food for associations, for accounts of Buddhism had entered into "every book of travels over these vast regions," and, though Hooker was not disposed to worship at the temples and sacred groves, "it was impossible to deny to the inscribed stones such a tribute as is commanded by the first glimpse of objects which have long been familiar to our minds, but not previously offered to our senses."[24]

But what intrigued Hooker most (as a botanist and biogeographer) was the abundant and luxuriant vegetation on the lower slopes of the Himalayas, which had a tropical character without actually being within tropical latitudes or tainted by many of the negative characteristics associated with the true tropics. Above this tropical zone of vegetation, at varying heights determined by aspect, altitude, and precipitation, were to be found the familiar temperate plants that he (and other British travelers of the period) readily associated with "home," to be eventually replaced by the alpine flora of the high mountainsides and passes. In terms of broad botanical classification, Hooker was able to map this tripartite distinction very precisely across a complex landscape of high mountains, lateral ridges, and deep ravines, identifying and naming characteristically tropical, temperate, and alpine genera and species.[25] Thus the presence of palms, figs, and bananas signified the domain of the tropics even when they reached heights of seven thousand feet in some of Sikkim's sheltered valleys, while oaks, yews, and birches were similarly eloquent indicators of temperate vegetation. The glorious rhododendrons, many of which Hooker collected and identified for the first time, lay in the upper temperate zone (though a few stunted specimens straggled into more alpine regions), and these were among the main botanical "finds" of his journey, but they were also central to his envisaging of Himalayan vegetation (fig. 8.4). "In the months of April and May," he wrote, "when the magnolias and rhododendrons are in blossom, the gorgeous vegetation is, in some respects, not to be surpassed by anything in the tropics." He had to admit, though, that at such altitudes the effect of this grand floral display was often "much marred by the prevailing gloom of the weather."[26] In describing the Himalayan flora, Hooker strove not only to collect, identify, and ultimately present, for pictorial depiction, individual plant specimens or to fit them into the broad typology of tropical, temperate, and alpine (or arctic) species but also to use them as a guide to typifying the vegetation of a given landscape—such as the characteristic firs and rhododendrons of the high temperate zone.[27] In this process of identification and typification, he extensively employed a language of aesthetics (of

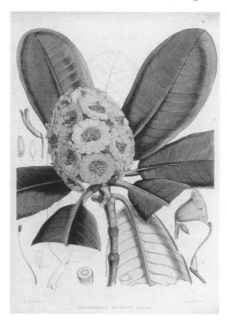

FIGURE 8.4 *R. hodgsoni,* drawn by J. D. Hooker and engraved by Walter Hood Fitch, from J. D. Hooker, *The Rhododendrons of Sikkim-Himalaya* (London, 1849–51) (by permission of the British Library)

color, form, and mass) as well as the more technical vocabulary of botanical science to enable him to recall (and the reader to visualize) the nature and diversity of the plants and scenery that greeted the traveler's eye. Thus, the passage on magnolias and rhododendrons just cited continues:

> The white-flowered magnolia (*M. excelsa, Wall.*), forms a predominant tree at 7,000 to 8,000 feet; and in 1848 it blossomed so profusely, that the forests on the broad flanks of Sinchul [near Darjeeling], and other mountains of that elevation, appeared as if sprinkled with snow. . . . In the same woods the scarlet rhododendron (*R. arboreum*) is very scarce, and is out-vied by the great *R. argenteum,* which grows as a tree forty feet high, with magnificent leaves twelve to fifteen inches long, deep green, wrinkled above and silvery below. . . . I know nothing of the kind that exceeds in beauty the flowering branch of *R. argenteum,* with its wide spreading foliage and glorious mass of flowers.[28]

Nature did not always obey the typifying urge of the experienced naturalist. At times the tropical, temperate, and alpine, instead of being poles apart, could be seen simultaneously from a single vantage point or were found mixed promiscuously together in jungles that were, botanically speaking, both tropical and temperate at the same time. On an early excursion from Darjeeling, Hooker observed of figs that "one species of this very tropical genus" ascended to almost nine thousand feet on the outer ranges of Sikkim.[29] In the Tambur Valley a few months later, he could look down from the "savage grandeur" of the black mountains overhead, their "rugged, precipitous faces . . . streaked with snow," to the tips of silver firs below, "contrasting strongly with the tropical luxuriance around."[30] Yet this ecological complexity and blurring of basic botanical divisions did not cause Hooker to abandon the idea that there were distinctively tropical, temperate, and alpine plants, though he did challenge existing biogeographical conventions by showing that in the Himalayas even temperate oaks could be "very tropical plants."[31] In both his notes, written en route or at his nightly bivouac, and in the published version of his journals, he held onto the value of the analytical and associational distinction between tropical and temperate vegetation.

The repeated contrast between "the tropics" and "the temperate zone" was not confined to the plant world or descriptive topography. It was also evident when Hooker shifted from botany to ethnography and tried to describe the inhabitants of the different climatic and topographical zones and their distinctive characteristics. If the sticky, sickly plains of India suited the "servile" and "indolent" Bengalis (as he saw them) or the "primitive" tribes of the malarial Tarai, so the temperate area around Darjeeling not only was inhabited by "English-looking," "European" plants but was also, appropriately, home to fresh-faced English children, who were so markedly different in their appearance from the sickly, yellow-faced infants Hooker had observed in Ceylon or in the Indian lowlands. The temperate belt of the Himalayas thus constituted an environmental norm from which both the torrid plains and the frozen high plateaus appeared, aberrantly, to diverge. Like Hodgson, Hooker began to see this part of the Himalayas as suited by nature to European colonization or, at the least, as a sanctuary "where the health of Europeans may be recruited by a more temperate climate."[32] The high alpine region, by contrast, though relatively disease-free, was too cold and barren to be of use for European settlement and was fit only for the warlike Gurkhas or the "quarrelsome, cowardly, and cruel" Bhutanese.[33]

Thus, for Hooker, as for many of his naturalist contemporaries, the idea of the tropics operated at three levels of perception and articulation. At the most general level, the tropics were an idea already present in Europe and North America. Since the voyages of Bougainville and Cook, educated Euro-Americans had been familiar with a literature of travel and natural history writing, along with a host of literary tropes and visual representations, in which the tropics were presented in both their Edenic and pestilential forms. Hooker did not need to invent the tropics: they already existed in his personal repertoire of images, allusion, and associations and

PLATE 1. William Hodges, *A View Taken [in the] Bay of Oaite peha Otaheite or Tahiti Revisited*, 1776, oil on canvas (National Maritime Museum, Ministry of Defence Art Collection, London)

PLATE 2. Johann Moritz Rugendas, *Tree in the Brazilian Forest,* 1830, oil on canvas (Stiftung Preussische Schlösser und Gärten, Berlin-Brandenburg; photograph by Roland Handrick)

PLATE 3. Frederic Edwin Church, *The Heart of the Andes*, 1859, oil on canvas (Metropolitan Museum of Art, New York, bequest of Margaret E. Dows)

PLATE 4. William Burchell, *Hermit Crab,* n.d., *St. Helena Plants,* no. 52 (Royal Botanic Gardens, Kew Archives)

PLATE 5. William Burchell, *Group of Plantains from Nature,* 20 February 1807, *St. Helena Plants,* no. 49 (Royal Botanic Gardens, Kew Archives)

PLATE 6. William Burchell, *Sample Colour Sheet,* Catalogus Geographicus Plantarum Brasiliae Tropicae, 1825–30 (Royal Botanic Gardens, Kew Archives)

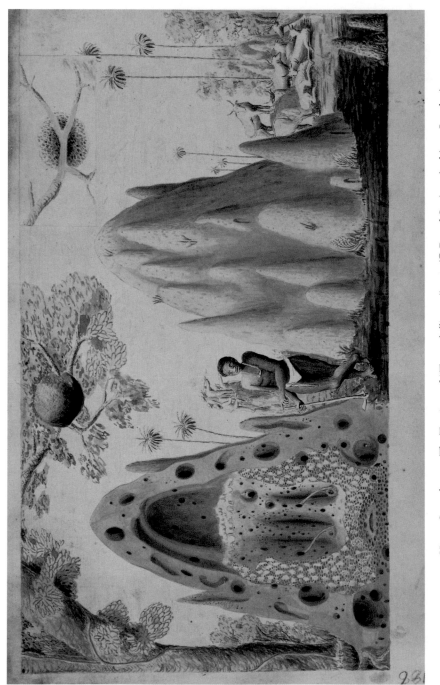

PLATE 7. Henry Smeathman, *The Territory of "Termes bellicosus,"* 1781 (© Royal Society Archives, London)

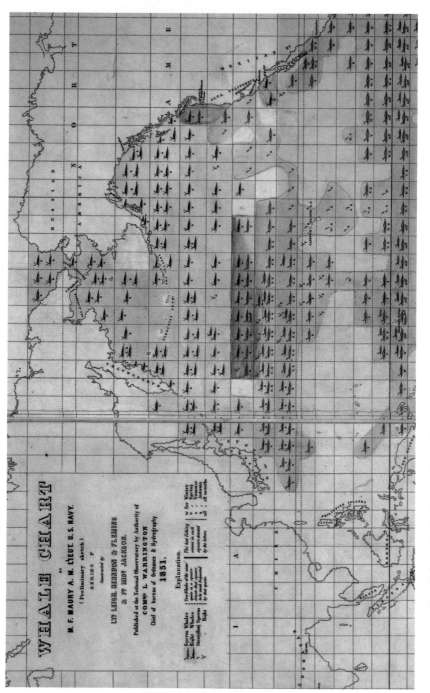

PLATE 8. Detail from M. F. Maury, *Whale Chart of the Pacific (Preliminary Sketch)*, 1851 (Library of Congress, Washington, DC)

PLATE 9. Aspects of Nature in Different Latitudes, from Arnold Guyot, *Physical Geography* (New York and Chicago: Ivison, Blakeman, Taylor & Co., 1885)

in the references ·(as to familiar paintings and novels) he could make whether in writing to his friends and family or for a wider public. To the popular visualization of the tropics, Hooker's most striking contribution was the image of rhododendrons and magnolias blooming amid subtropical vegetation beneath the snowcapped Himalayas.

But at a second level, the idea of the tropics required a more precise signification. This drew on medicine and the emergent sciences of the late eighteenth and early nineteenth centuries (botany, zoology, ethnology, geography, and meteorology), as it drew also on the experience of the expatriate resident and the tropical traveler, to create an interconnected world of meaning. Artistic and literary associations might still provide useful reference points, but the materiality and specificity of the tropics was here much more in evidence, and in his descriptive writings Hooker tried to refine his earlier expectations of the tropics and to add to the store of popular and scientific literature his more complex and in some ways contradictory impressions of the tropics. Emphasis might be placed on one aspect of the tropics—the vegetation, say, or the climate—but the ensemble was always more than the sum of its constituent parts. Neither climatic conditions, such as heat, humidity, and violent storms, nor a specific type of tree and flower was sufficient in itself to define the experience of the tropics and capture their scenic, sensual, and scientific complexity. Building on the work of Humboldt, Darwin and others, Hooker tried to show how the combination and interdependence of many different elements made the tropics distinct from the temperate zone but also enabled informed comparisons to be made between one part of the tropical world and another.

And finally, Hooker as a botanist could use the idea of the tropics in an even more precise way. Some plant species and genera were considered on the basis of extensive observation and comparison to be characteristic of the tropics as others were of temperate or alpine regions. As Hooker demonstrated in the Himalayas, tropical species might in fact be found outside the tropics as defined by lines of latitude and even in a secluded valley several thousand feet above sea level, but this did not necessarily affect their underlying identification. Used thus, the term "tropical" served as a botanical shorthand, indicative of the home territory of a particular plant genus or the part of the globe where it flourished best or was present in the greatest variety of forms. In an age in which the identification and transfer of "useful" plants and the discovery of new ornamental trees and flowers were central to economic botany and gardening, information about a plant's original habitat could prove of great practical importance. In his detailed botanical investigations, as in his more general scenic descriptions, Hooker helped establish the ecological context of particular plant species and contributed substantially to the more complex scientific understanding of the tropics.

FROM THE SEA TO THE SNOWS

For naturalists as well as painters, the sea was redolent with meaning and associations. Like Charles Darwin and T. H. Huxley, Hooker spent a significant part of his scientific apprenticeship at sea. From 1839 to 1843 he served as on-board naturalist on Ross's expedition to the Antarctic, Tasmania, and New Zealand. Not only did the four-year voyage (and his well-received botanic reports) establish Hooker as an expert on the temperate and arctic flora of the southern oceans, it also trained him (rather as Bernard Smith has argued with respect to painters in *European Vision and the South Pacific;* see also chap. 2 in this volume) in the empirical observation of sea, land, and sky.[34] These habits and techniques persisted throughout Hooker's Indian journey, even when he was many hundreds of miles from the sea. Several times a day he took temperature readings, checked his barometers, and made extensive notes on the weather, recording particular cloud formations and distinctive atmospheric conditions.[35] Trained to the ways of the sea and accustomed to studying coastal features from on-board ship, he read the landscape of Himalayan India as if it were literally, as well as metaphorically, the shoreline of a now vanished sea. This was a time when British geologists like Charles Lyell had learned to recognize the wave-cut terraces and fossil traces left by the rise and fall of prehistoric sea levels, and Hooker was keen to utilize this knowledge in his exploration of the Himalayas to advance discussion of their origins and form. Not only did the south-facing declivity of the Himalayas look like the coastal fjords of Scotland, Norway, and New Zealand, but, he argued, it actually was a relic of the time when the Himalayas were washed by an ancient sea, before the seabed was itself uplifted to form part of the long mountain chain.[36] In making this essentially scientific observation, Hooker was also able to deploy a series of literary embellishments and allusions that helped to enrich his narrative of botanical exploration and make it more attractive to his readers.

As perhaps befitted a young and aspiring professional botanist, Hooker's earlier scientific writings had been matter-of-fact, unadorned by attempts at more evocative description.[37] But in preparing his journals for publication, and still smarting from the damning review in the *Athenaeum,* he sought to develop a prose style that would be sufficiently rich in "pictorial power" to engage a wide readership while at the same time being sufficiently technical to communicate his scientific understanding of how topography, geology, vegetation, and climate worked together to build a single dynamic entity. He did this by employing a romantic style of writing, which partly drew on literary and pictorial sources but was also influenced by the travel narratives of Humboldt and Darwin, whose account *Voyage of the "Beagle"* was first published in 1839.[38] In this romantic style (and without a professional artist to depict the scene for him), Hooker paid particular attention to the use of language to evoke color, form, and general atmosphere and he situated the observer-naturalist squarely at the center of the landscape he was simultaneously experiencing and describing. At times he invoked a kind of romantic reverie to suggest the intensity of his emotional experience and the depth of his meditations on natural phenomena. Perhaps these stylistic devices were all the more important to Hooker given the striking secularity of his writing. Since there is no obvious moral or religious message in the *Himalayan Journals,* no evident intention to interpret nature as God's handiwork, an intensity of human perception and scientific understanding was necessary to give his narrative structure and purpose.

The following passage from the *Journals* illustrates some of these ambitions and effects. Hooker here recounts his first ascent from the plains into the Himalayan foothills. Looking back over the "sea-like expanse" of the plains below, he is presented with a scene that elicits a Humboldtian perception of a vast, complex, but ultimately integrated and harmonious universe.

> In the distance, the course of the Teesta and Cosi, the great drainers of the Himalayas, and the recipients of innumerable smaller rills, are with difficulty traced at this, the dry season. The ocean-like appearance of this southern view is even more conspicuous in the heavens than on the land, the clouds arranging themselves after a singular seascape fashion. Endless strata run in parallel ribbons over the extreme horizon; above these scattered cumuli, also in horizontal lines, are dotted against a clear grey sky, which gradually, as the eye is lifted, passes into a deep cloudless vault, continuously clear to the zenith; there the cumuli, in white fleecy masses, again appear; till, in the northern celestial hemisphere, they thicken and assume the leaden hue of nimbi, discharging their moisture on the dark forest-clad hills around. The breezes are south-easterly, bringing that vapor from the Indian Ocean, which is rarefied and suspended aloft over the heated plains, but condensed into a drizzle when it strikes the cooler flanks of the hills, and into heavy rain when it meets their still colder summits. Upon what a gigantic scale does nature here operate! Vapors, raised from an ocean whose nearest shore is more than 400 miles distant, are safely transported without the loss of one drop of water to support the rank luxuriance of this far distant region. This and other offices fulfilled, the waste waters are returned, by the Cosi and Teesta, to the ocean, again exhaled, exported, expended, re-collected, and returned.[39]

FIGURE 8.5 *Wallanchoon Village,* from J. D. Hooker, *Himalayan Journals* (London, 1854) (by permission of the British Library)

A later scenic description, carefully worked up by Hooker from his earlier notes, finds him at twilight on a ridge high in the Himalayas. As the daylight begins to fade,

> a sudden chill succeeded, and mists rapidly formed immediately below me in little isolated clouds, which coalesced and spread out like a heaving and rolling sea, leaving nothing above its surface but the ridges and spurs of the adjacent mountains. These rose like capes, promontories, and islands of the darkest leaden hue, bristling with pines, and advancing boldly into the snowy white ocean, or started from its bed in the strongest relief. As darkness came on, and the stars arose, a light fog gathered round me, and I quitted with reluctance one of the most impressive and magic scenes I ever beheld.[40]

During his Himalayan excursions, Hooker was engaged in a series of parallel botanical, geological, and topographical exercises. He devoted particular care to constructing an accurate map of Sikkim, a region previously little known to Europeans. For the government of India, anxious about its northern frontier and relations with China, this was one of the most valued outcomes of Hooker's expeditions.[41] The published version of the journals also contained a number of black-and-white engravings (as well as a few in color) based on the sketches he made at the time. Some of these were used to illustrate the ethnography of the region and Bud-

FIGURE 8.6 *Tambur River at the Lower Limit of Pines,* from J. D. Hooker, *Himalayan Journals* (London, 1854) (by permission of the British Library)

dhist worship, but most were depictions of Himalayan scenery, designed, with few artistic pretensions, to represent the main features of the landscape and vegetation (figs. 8.5 and 8.6). They do considerably less, however, than Hooker's elaborate prose descriptions to capture the grandeur of the mountains and bleakness of the passes and can do little to articulate the complex train of scientific ideas and visual associations that informs his prose narrative.[42] It is significant that it was these elaborate (and perhaps to our minds excessively ornate) descriptive passages that attracted the most attention and praise from reviewers. Six years after his letters had met with disdain in the journal, a lengthy and enthusiastic review in the *Athenaeum* commended the "artistic spirit" with which Hooker "reproduced some of the grander features of the scenery through which he passed."[43]

One of the passages singled out for particular commendation related Hooker's visit to the Choonjerma Pass on the Nepal-Tibet border in December 1848. Here again Hooker evoked the idea of clouds forming a "misty ocean" at his feet, but now the "sea of mist" was turned into "a blaze of the ruddiest coppery hue" by the rays of the setting sun.

> As the luminary was vanishing, the whole horizon glowed like copper run from a smelting furnace, and when it had quite disappeared, the little inequalities of the ragged edges of the mist were lighted up and shone like a row of volcanoes in the far distance. I have never before or since seen anything, which for sublimity, beauty, and marvelous effects, could compare with what I gazed on that evening from Choonjerma pass. In some of Turner's pictures I have recognized similar effects, caught and fixed by a marvelous effort of genius; such are the fleeting hues over the ice in his "Whalers," and the ruddy fire in his "Wind, Steam, and Rain," which one almost fears to touch. Dissolving views give some idea of the magic creation and dispersion of the effects, but any combination of science and art can no more recall the scene, than it can the feelings of awe that crept over me, during the hour I spent in solitude amongst these stupendous mountains.[44]

CONCLUSION

Despite the scientific importance of his travels and the calculated use of verbal effects to communicate a personalized sense of the tropics and Himalayas to a wide readership, Hooker's *Himalayan Journals* were no more than a partial success. Certainly, when it appeared in 1854 the work was well received by the scientific community: Darwin, to whom it was dedicated, was particularly enthusiastic. It was, he said on completing the first volume, a book that would become "a standard," not so much because it contained "real solid matter" but because it gave "a picture of the whole country,—one can feel that one has seen it . . . & one *realises* all the great Physical features."[45] But others found the work over-long and rather awkwardly divided between Hooker's personal adventures and his scientific observations. "A narrative of nearly eight hundred pages," complained one reviewer in the *Spectator*, "requires to be lightened by every artifice of composition."[46]

Hooker's *Himalayan Journals* demonstrate, nonetheless, the extent to which a scientific writer in the early Victorian age was prepared to go, and the range of devices he was prepared to employ, in order to render his tropical travels and natural history narrative accessible to a wider public. Apart from maps and engravings, the printed word (and the multiplicity of images and allusions it could be used to conjure up) was consciously deployed to engage, entertain, and inform. At the same time, for reasons of both scientific knowledge and popular appreciation, Hooker made extensive use of the idea of the tropics. It was part of his achievement as a scientist to bring India and the Himalayas more securely within the conceptual domain of the tropics. He did so not only (as his standing as a professional botanist required) by close and careful identification of plant species and their distribution but also by evoking (after the manner of Humboldt and Darwin) a wider emotional and experiential understanding of the tropics, painting verbal pictures and summoning up mental associations that would help him and his readers to typify the tropics. As a trained naturalist, he well understood the complex ecological relationship between tropical and temperate plant life, but he still recognized the heuristic force of that distinction, especially in situating the exotic tropics relative to the more familiar norm of the temperate zone.

⋆ 9 ⋆

Eyeing Samoa: People, Places, and Spaces in Photographs of the Late Nineteenth and Early Twentieth Centuries

LEONARD BELL

During the late nineteenth and early twentieth centuries, Samoa—or rather certain notions about and views of Samoa—became well known in Western Europe and North America through the medium of photography. Large quantities of images were produced both by resident European professionals in Samoa and by visiting photographers. This level of production initially created and then sustained a market for images of Samoa—a body of images that effectively came to stand more generally for the tropical. A recent exhibition of "colonial photographs of Samoa, 1875–1925," shown in England, the United States and Germany, was titled Picturing Paradise (*Bilder aus dem Paradies*). These titles suggest the prevalence of certain types of imagery in the representation of Samoa, reflecting a certain vision of Samoa from the outside—in Europe and America—during the colonial period and after: the place seen or imagined as a paradigmatic idyllic, "South Seas" tropical island. However, while there may have been plenty of photographs that fueled such a view, there were also other kinds of photographs made in Samoa by people living and working there, attuned to the particularities of specific local experiences rather than to generalities about the tropical. These images suggest a more problematic terrain, which resists any straightforward or totalizing characterization of the tropical, whether in terms of paradisiacal and prodigious nature or its opposite, the degenerate and disease-ridden. As represented photographically, the tropical, or at least tropical Samoa, can be seen as a set of socially interactive spaces more complexly inflected than the standard view has allowed.

FIGURE 9.1 John Davis, "Robert Louis Stevenson with Family, Friends, and the Band of HMS *Katoomba,* at Vailima, Samoa," ca. 1893 (Alexander Turnbull Library, National Library of New Zealand, Wellington)

For a sense of this shifting ground, consider two photographs of the same place in Samoa by resident photographers: one from the early 1890s by John Davis, the other about ten years later by Alfred Tattersall, once Davis's assistant.[1] The first, in figure 9.1, shows a crowd of people, fairly close up, as if the photographer, too, is part of the event and space. The extraordinarily mixed group includes the famous writer, Robert Louis Stevenson and his wife, settlers in Samoa, local inhabitants, both Samoan and of Samoan and European lineage, and sailors from a visiting naval ship, HMS *Katoomba,* all crammed together on the veranda of Stevenson's house at Vailima. This was a most unconventional household, renowned for its frequent blurring of expected colonial hierarchies and boundaries. In respect of this, the American writer Henry Adams, visiting Stevenson in 1890, complained that Stevenson "does not know the differences be-tween people, and mixes them up in a fashion as grotesque as if they were characters in his New Arabian Nights."[2] Figure 9.2 shows another and very different view of the same veranda, then—circa 1905, after Steven-son's death and the dispersal of his family—the official residence of

FIGURE 9.2 Alfred Tattersall, "Dr. and Mrs. Solf at Vailima," ca. 1905 (Alexander Turnbull Library, National Library of New Zealand, Wellington)

Dr. Wilhelm Solf, the first German colonial governor of Samoa. Hierarchy and concomitant spatial divisions have apparently been firmly reestablished. This is a view from further back, with the photographer in his own separate place and space, and Dr. Solf, wife and baby, the ideal family image, dominate this colonized space. They are the focal point, the center of this world, with their Samoan attendants, in their subaltern costume, secondary staffage, subordinate figures—in keeping with Solf's vision of Samoa, fundamental to which was a belief in the notion of "pure" races or ethnicities, and with attempts to classify the population of Samoa in terms of fixed "racial" or ethnic categories.[3] There is no mixing of disparate bodies here.

Before considering further examples, location and contexts need to be established. Samoa is a group of small islands in the geographical tropics, more specifically, in Polynesia, in the South Pacific, about 2,500 kilometers northeast of New Zealand. The places and peoples there have been subject to many and different kinds of "eyeings" since first European contact. From the 1880s to the early years of the twentieth century, for complex geopolitical reasons, Samoa was subject to the intent scrutiny of the three so-called Great Powers—Britain, Germany, and the United States—

as well as a small settler colonial society, New Zealand. The latter was not, in fact, very powerful or comfortably settled, but its leaders could have large, at times grandiose, ambitions and visions. In the late nineteenth century, Samoa was not a settler colonial society like New Zealand or Australia, in the sense that those countries had had large-scale immigration of Europeans, who soon outnumbered the indigenous peoples and effectively "took over." There were, though, around four to five hundred European residents in Samoa in the 1890s, about half of whom lived in Apia, a port town that only came into existence with European contact. Apia was an autonomous municipality from 1879, and control of the town and residents' affairs was in European hands.[4] It had, though, a culturally and ethnically heterogeneous population, significant numbers of whom were of "mixed" parentage or ancestry, not readily classifiable in terms of single or fixed racial categories, and some of whom were inclined to be "troublesome."[5]

In the 1880s and 1890s, especially with the 1889 Berlin Treaty, the government of Samoa was effectively controlled in many respects, directly and indirectly, by the colonial powers. But Britain, Germany, and the United States were frequently at odds with another, jockeying for position and attempting to gain advantage for, or supposedly protect, the interests of their nationals there. It was a period, too, of complex and internecine rivalries, power struggles, and civil unrest within Samoan society, a very unstable situation, precipitated by European incursions and alienation of, and contestation over, land. The Great Powers used this instability as a pretext for intervention, supposedly to establish order and functioning modes of government, administration, and jurisdiction. Relationships between the various competing groups, European and Samoan, were constantly shifting. It was a confused and confusing state of affairs resisting simple or singular explanation—thus the title of a book on the machinations and maneuverings of the various "players": *The Samoan Tangle*.[6] Samoa, in short, was difficult to manage and often chaotic politically. The three European nations, though, aimed for fixity, order, regulation. To this end, the Berlin Treaty established a tri-dominium and imposed Malietoa Laupepa as king of Samoa, though most Samoans favored his rival, Mata'afa Iosefo.[7] That did not solve the Samoan "problem," which only achieved resolution, temporarily it turned out, with the colonial division of Samoa between Germany (which took the bulk of the island group, as Western Samoa) and the United States (the smaller eastern bit) in 1900.

Although it was a time of sociopolitical instability, conflict, and confusion, the increasingly prevalent popular image of Samoa in Europe and the United States was that of an exemplary tropical island, with accompanying attributes—hot (in several senses), luxuriant, verdant, welcoming, easy living, and with contented and attractive "natives": a model of the exotic picturesque. This image, disseminated and sustained primarily through photographs in various formats—*cartes de visite,* prints, reproductions in books and magazines, and postcards—was facilitated by the position of Samoa, in particular Apia, as a key node in trade and communications networks and routes between the western coast of North America and Australasia.[8] There was a ready mechanism for distribution and, thus, a wide circulation of photographic images of tropical nature and peoples—for example, beaches and coastal strips replete with palm trees (an essential sign of the tropical) and occasional canoes, waterfalls, forest scenes, "untouched" villages, "pure" Samoans in traditional dress engaged in their "manners and customs." Many foregrounded bodies, costume, and artifacts, the figures mostly young and good-looking, the women frequently bare-breasted and "comely"—often, stereotypical "dusky maidens." The pictorial lineage of such images can be traced back, ultimately, to paintings like John Webber's *Poedoea* (1777), the subject of which had been taken captive by Cook on his third Pacific voyage. The idea, and ideal, of the paradisiacal tropical island was, after all, embodied by the young, "available" female.

Photographs of this sort belong to a particular construction of the "eyeing" of Samoa as an objectifying, possessive scrutiny, of Samoa and Samoans as objects of a dominating European vision, rendered passive and compliant within a project of surveillance and control.[9] While recognizing that the camera has been used as an instrument of power and domination, it is not necessarily or only so, and in recent work, I and others have questioned the adequacy of this construction or mode of seeing in relation to visual representations of the colonial period.[10] As a number of writers have noted, the re-representation or preoccupation with a limited range of such stereotypical images and the viewing of them in terms of objectifying possession, in which the photographed subject is allegedly rendered passive and silent, can serve, ironically, to reinscribe and perpetuate the stereotypes and objectifying vision that the writers are supposedly exposing and critiquing.[11]

Rather than reviewing the standard, stereotypical photographs, in this chapter I consider a small number of less familiar images of Samoa—a dance scene or theatrical performance with spectators, an Apia street scene with a mixture of figures looking back at the photographer and viewer, a Samoan village scene with a foregrounded male figure, a group of armed Samoans confronting the viewer, a man sitting on a footbridge in a rural location. Most of these were produced in the same Apia studio, founded by John Davis, as many of the standard photographs, which served the image of Samoa as a model tropical island. But these examples involve more complex, nuanced, perhaps even critically reflexive eyeings of Samoa, its inhabitants—Samoan, European, and those neither simply nor purely European or Samoan—and their negotiations and contestations over the occupancy and use of space. These images are less amenable to straightforward and reductive explanation. They can be seen, rather, as sites in which a variety of relationships and viewing positions coexist, sometimes uncomfortably. Running counter to any notion of an internally consistent and univocal "body" or archive of Samoan images, they provide a counterpoint to stereotypical visions of tropical Samoa. These images—tropical, in another sense—"turn" figuratively. They offer possibilities of viewing in terms of metaphor, parody, and deviations from the normative. It becomes a matter of considering what can happen in, or be done with, the space of the photograph—what photographs can do.

John Davis (1831–1903), described as "one of the oddities and most liberal men of the Beach," came from Sydney, where it is likely that he had been manager of the Sydney Photographic Company in the early 1870s.[12] He founded the first commercial photographic studio in Apia within the same decade. He produced a large and varied body of work, assisted from 1887 by Alfred Tattersall, a New Zealander who bought Davis's studio and photographs in 1895.[13] Some of these were later marketed under Tattersall's name. Establishing precise authorship and dates of some photographs can be difficult, though my recent research indicates that Davis was the prime executor of those photographs from his studio that are datable up to its sale. Photographs from the Davis studio had wide circulation and a variety of uses, both inside and outside Samoa. One and the same photograph could, for instance, appear in the substantial private photo albums (now in a public collection) of Sir Thomas Berry Cusack-Smith, the British Consul in Apia from 1890 to 1898, and in the German doctor, Augustin Kramer's two-volume scientific treatise, *Die Samoa Inseln* (1901–3), the first ethnographic and ethnological study of Samoan people and culture.[14] Kramer used photographs extensively in this work (nearly two hundred) and commended Davis, and another resident commercial photographer (Thomas Andrew), for the "excellence" of their work.[15] It was, of course, not uncommon, for photographs originally made for commercial, exotic picturesque, touristic, and other purposes to be incorporated into ethnographic and anthropological studies and bought by museums as supposed records of the customs, physical appearances, and material culture of particular ethnic groups.[16]

Much of Davis's (and Tattersall's) output was routine, catering generally to standard consumer tastes, the types most in demand: portraits, "pure" natives in traditional costume, their "exotic" manners and cus-

toms, and scenic images, for example, as well as some that referenced top-
ical and newsworthy events or showed the work on, and products of, European plantations. This has led to the output of their studio being characterized by a leading historian of Samoan photography as "combining
themes of primitive exoticism with an emphasis on the positive influence
of western development; a pattern that occurs with great regularity in both
the photographs and the verbal rhetoric of the period."[17] This view is reiterated in a recent book on colonial photographs and exhibitions, in
which Davis and Tattersall are portrayed as photographers who did not
"move beyond the production of colonial stereotypes."[18] Not so, I'll
argue.

Whether, or to what extent, such works served colonialist agendas, insofar as they could be fitted into systems of classification and categoriza-
tion of peoples and places amenable to the dominance of colonial powers,
is not my concern here. Rather, I will view a selection of Davis Studio
products from another perspective, emphasizing different aspects of the
relationships involved—between photographers, viewers, and photographed subjects and the various sorts of spaces these interacting parties
inhabited. These photographs present spaces in which the complexities
and ambiguities of interrelationships between outsiders and insiders—and
those neither just one nor the other—can be explored and questioned.
This approach may provide a sense of actual practices otherwise masked
by standard visions of tropicality. Thus, while not losing sight of the wider
sociopolitical and cultural contexts, my focus is on specific photographs,
with close attention to elements of detail. These aspects have been largely
overlooked in many writings on Samoan photographs.[19]

The first of the photographs I consider is the dance: "Siva" (fig. 9.3)
belonged to a series of prints (including "The Reading of the 1889 Berlin
Treaty to Malietoa Laupepa," another staged event) available to buyers in
the 1890s.[20] "Siva" shows an ethnically mixed group of people in a semi-
circle in near mid-ground spanning the image. Most, though not all, are
watching a performance in the far mid-ground, a performance that,
though titled "Siva" (a dance form), looks more like a theatrical tableau.
Several of the audience—a European woman on the far left, and some on
the right, including Samoans—are not viewing the show but are looking
back out of the picture at what is another performance: that of the photographer taking the photograph. He can be located by the convergence
of those outward looks, just out of the space of the photograph. A complicated interplay of lookings and viewings is in play. The staging and performance of the photograph, then, is as much the subject of the photograph as the ostensible subjects: the theatrical performance, an aspect of

FIGURE 9.3 John Davis, "Siva or Samoan Dance," ca. 1895

the indigenous culture staged for the occasion, and the viewing of it by an audience.

This combination of elements was most unusual, radically different from the standard. The great majority of photographs of *siva,* a popular subject and prime marker of an exotic "Samoanness," show just the Samoan dancers, with the focus on movement, gesture, and costume, on bodies and the activity as a "sight," as objectively rendered documentations of a culturally distinctive activity. In Davis's "Siva," in contrast, there is no prime focus on the performing bodies. While there are a few later photographs of *siva* that include, only peripherally or marginally, spectators, there are no others, to my knowledge, that place the audience and their viewing at center stage, foregrounding spectatorship itself.[21] Nor are there other photographs that draw attention to the act of making the image, establishing thus the photographer as both orchestrator and part of the event—as one who is copresent (if not literally in the picture), not separate from it and neutrally documenting what would exist without him.[22]

Rather than just delivering up a sight, then, this photograph raises questions about specific social relationships and conditions of spectatorship. For this is a photograph that unconventionally would have made the European viewer aware that he or she was a viewer and, also, that this enactment, or reenactment, of a supposedly customary practice was a staging of

"Samoa." The photograph shows that the space in which the represented event took place was socially produced, not "natural." The typical South Seas sight and photo image, then, was represented as an act performed for viewers: the exotic tropics manufactured.

To make a speculative aside, there could be more to this picture. While it is difficult to make out clearly what the performing figures are doing, they are not actually engaged in dance. Rather their poses, gestures, and positions in relation to one another suggest a comic or caricatural act. That could have been the case, since there is a rich social history of clowning in Samoa, of the performance of comic sketches, both scripted and rehearsed and spontaneous and unplanned.[23] This cultural phenomenon was a complex and elusive one, difficult to pin down, since the comic sketch, which was usually associated with dance activities performed outdoors and casually mounted with little or no use of costumes and props (as in the photograph), provided the only public arena in which criticism of figures and institutions of authority could be voiced—of people who otherwise might be among the audience. This could include sending up Europeans of power and authority, the criticism being veiled by irony, parody, and the "lightness" of entertainment, since in Samoan society authority figures were traditionally accorded respect.[24] (Interestingly, in this context, the central European pair in the photograph, in particular the female figure, even if back to the viewer, resemble strongly the then-representatives of British authority, Consul Cusack-Smith and his wife, Winifred, an artist. There are numerous photographs of this couple, who, coincidentally, organized, staged, and had photographed their own theatrical, usually comedy, performances).[25] The Samoan comedy sketch involved breaking the frame of social conventions and the expression of views that could not otherwise be publicly expressed. Likewise with "Siva," the conventional representational frame for photographs of "traditional" Samoa was broken, allowing reflection on the manufacture of such tropical views. The authority of the camera, then, as an instrument of attempted domination, could thus be brought into question.[26]

Such a proposition might seem to be granting too much sophistication, and more especially a degree of self-criticism, to the photographer. However, the settler photographers could play games with their audiences. Consider this story about the working practice of the other professional photographer active in Apia in the 1890s, Thomas Andrew, as recounted in a manuscript memoir by a European settler assistant, George Westbrook.[27] It is an account of a knowing and deliberately misleading staging of a visual cliché, a commonplace "tropical vision." Andrew's "star picture" featured a young Samoan woman, seminude, standing beneath a waterfall—a variant of that longstanding popular image in Western visual

FIGURE 9.4 Alfred Tattersall, "Soldiers Making up the American Guard," ca. 1899 (Alexander Turnbull Library, National Library of New Zealand, Wellington)

culture, the bathing nude.[28] The girl was posed against an illusionistically painted screen and a pile of rocks on an unseen veranda; the liquid running down the rocks, a mixture of condensed milk and actual water, poured from oil drums hung out of sight of the camera eye. The viewer was hoaxed. What you see is not what you get. The resident or settler photographers well knew that Samoa did not correspond to the idea or fantasy of the idyllic or unspoiled tropical island, to which many of their images otherwise catered. To make the link with "Siva," the photograph can be viewed, then, as a discrepant image that allows a sense of differences in experiences and vision among ethnically and socially diverse figures occupying the same physical space.

Now to photographs that present the streets of Apia and their various occupations: figure 9.4 shows an American guard strategically placed in a public space during a period of civil unrest. In comparison, consider Davis's "Street in Apia" from the late 1880s (fig. 9.5), which at first glance may appear to be a straightforward descriptive image. As in the previous image, the street is lined with as variety of verandaed colonial buildings and picket fences and narrows to a distant point, as if the composition had been consciously modeled on that classic spatial schema in which picto-

FIGURE 9.5 John Davis, "Street in Apia," ca. 1886

rial elements are organized in terms of linear perspective with a vanishing point. Twelve standing figures of various sizes and ages, predominantly Samoan in dress (lavalava, lei) and appearance (though one child looks part Chinese) are disposed at various points from near mid-ground to background across this receding street. In those pre–snapshot days and to avoid blurring, the subjects would need to have been placed and positioned by the photographer, if not posed by him. No European figures are visible, though the buildings establish European presence. And the sign on the right front building reads "J. Davis Photographer," effectively signaling his participation in the event and the relationships represented. That Davis was both inside and outside this space was apt, given the existential and psychological space frequently occupied by settler colonials—part and not part of the place, just as Apia was both inside and outside Samoa proper, a contestable site, in which people of various ethnicities and cultures encountered one another, sometimes mixing, sometimes clashing. While Apia may have been under European control, the place of Samoans in it—displaced but still there—could be a source of tension and resentment.[29]

The photograph in figure 9.5, a carefully composed scene, could be seen as another performance. The street is a public space; the houses into which viewers cannot see are private spaces, though their facades mark the point of meeting between the public and the private.[30] Pointedly, the social and physical boundaries between these different spaces are signaled by the picket fences. The architecture also functions to frame and enclose the street and public space, a living space, too, concentrating the eye on the specifics of the scene, in which the Samoan figures stand, face to face with the photographer and, by extension, the photograph's viewers. Apia was not just a place where different people coexisted but one where they could mix in unclassifiable ways. At the same time, some clear and enforced demarcations remained, as this photograph implies. Hotels, for instance, were not supposed to sell liquor to Samoans. They were out of bounds.

I noted above that the structuring of space in this image echoes the Renaissance perspectival model. That organizational system positions the viewer in relation to what is pictured and is geared to coherent and rational order, to the harmonious integration of parts with the whole, to the control of space for the outside viewer and user. Here, though, in the pictorially composed space of Apia, there are countervailing elements, which bring into question the spatial and organizational dynamic of control and order. Those are the non-European figures, or some of them, not simply because of their presence but as a result of how they are positioned and posed and of how they look in their encounter with the photographer and European viewer. While these Samoan figures are within the enclosure of the street, the forward adult males, at least, do not appear to be subordinated by an overseeing eye, though they are, of course, subject to being looked at. Their poses are straight on, forthright, confident, their returned looks direct and assertive (though some of the children are tentative-looking). They function as partial stoppages in the receding space of the street. The foremost group spans the street, fence-like, arresting the gaze, serving to interrupt a sense of an uncomplicated unitary space. The space, then, which on the face of it might seem the subject of a unifying vision in which all elements are unproblematically part of a controlled space, can also be seen in terms of discrepant social relationships. In the photograph, several separate physical areas sharply abut. Angularities, sharp edges, pointed accents abound, while the figures obstruct straightforward visual entry to, as well as passage through, the space of the street. The viewer is presented with differing sociocultural and psychological spaces, which are simultaneously brought together and separated. What viewers can see here, and in other photographs of Apia streets, are the social and political implications of the regulated yet heterogeneous space that was Apia.[31] There are other photographs of Apia streets (such as fig. 9.4), with the same or similar compositional structure, in which various and different groups of people, similarly positioned facing the camera and viewer, occupy the public space. Cumulatively, they comprise a narrative of the street—as a political site and as a manifestation of the shifting fortunes of the contending "players."

FIGURE 9.6 John Davis, "Scene at Matafele, Apia," late 1880s–early 1890s
(Alexander Turnbull Library, National Library of New Zealand, Wellington)

What was implicit in figure 9.5 is made explicit in the photograph
of another Apia street with socially and ethnically diverse inhabitants
(fig. 9.6).[32] Here, too, an element in the photograph—the post office sign
in the right foreground—points to the photographer himself, as Davis was
also the postmaster. Thus the copresence of the photographer in the space
and event represented is again signaled, now in a space represented un-
ambiguously as fiercely contested, given those dominant foreground
Samoan figures squaring up with rifles. As such, the photograph references the periodic outbreaks of violent unrest in Apia. And here the
boundaries of the hotel appear to have been breached, or partially so, since
the ground-floor door remains barricaded.

These photographs provide a glimpse of the instability and uneasiness
of the constantly changing social and political relationships in Apia and
Samoa in the 1890s, when attempts by the colonial powers to impose certain orders on Samoan places and peoples were forcibly resisted. These
tropical views are strikingly contrary to the standard, whether desired or
believed-in, of docile, compliant Samoan bodies and peaceful, inviting
Edenic scenes and locations—those essential constituents of an exotic

FIGURE 9.7 John Davis, "Group of Men and Women Armed with Traditional Weapons," early 1890s (Alexander Turnbull Library, National Library of New Zealand, Wellington)

South Seas vision and a popular image of the tropical (which lives on to the present in touristic representations). And the Samoan photo archive of the colonial period has generally been presented as if both visiting and resident photographers had little or no recognition of Samoan experiences and views, in particular, those that may have been in conflict with colonial projects.

In this respect, consider the photograph of a group of armed, no-nonsense-looking Samoans, both male and female, confronting the photographer and the viewer (fig. 9.7).[33] The encounter has the appearance of a standoff, but one can assume that the photographer, in order to take the photograph, would have needed the assent of these people. Their demeanor and assembly can be related to an episode of civil unrest in the period. Envisioned here is Samoan space, which, while obviously represented by the camera, has not been "captured" by it. It is space in which people stand their ground, their returned looks not welcoming. It is a culturally and psychologically interior space, which includes no signs of European penetration, unless one subscribes to the notion that the very act of photography or visual representation in itself "penetrates."[34] The presentation of the two female figures in this image is startlingly unconventional. They are not the usual "dusky maidens" but tough, solid characters, right in the foreground, dominant figures in this engagement and others that are implied. This presence was likely to refer to their status as *taupou,* daughters of high chiefs, both protected and protectors, who carried out important ceremonial roles and embodied the social and spiritual well-being of their group. One of their traditional functions was to lead men into battle.[35] In contrast to this image, there were many photographs made for the global market, titled *Taupou,* in which conventionally beautiful young women were displayed, usually with eye-catching costume and accompanying artifacts, in studio spaces or against leafy backdrops, which close off the space they inhabit—"captive" space.[36] Compare that with the open space of figure 9.7, a space in which boundaries are not clearly demarcated and that (given the cropped figures on either side) appears to extend beyond the photograph. It is an uncontrolled rather than an ordered space, in relation to which it should be noted that during neither the period of the tri-dominium nor that of the German colonial administration (1900–1914) was full control held over the Samoan population. Their potential for armed resistance remained considerable.

The final two photographs to be discussed here suggest the coexistence of different views within a single image, as well as the body of images as a whole. They draw attention to the possibility of the visual representation of relationships in a tropical site as multiple and contradictory, involving conflict over the control of space. Figure 9.8 shows a semiopen space in a forest. This might appear at first sight to be a luxuriant glade in the tropics, in which a Samoan man relaxes on a footbridge over a river or creek, a typical image of the paradisiacal, of oneness with nature, that the viewers are invited to enjoy, the bridge linking their space with his. However, the photographer (Davis or Tattersall) has complicated this scene by the inclusion of the sign on the tree in the upper right, behind the figure. This sign marks the boundary of a European plantation and announces in Samoan that entry is not permitted, except to plantation workers. Such workers in the colonial period were almost all Melanesian and Chinese, imported for the purpose. The man on this side of the sign represents he-who-cannot-enter the space beyond, now alienated from the Samoans— this in a period in which contention over the extent of European claims to land was the central issue in the political tensions between Samoans and the colonizing powers.[37] The photograph, then, is not quite what it initially seems. It is double-edged, turning on a fundamental irony. The seemingly stereotypical tropical view, bounteous nature, is undermined by that sign. Once its meaning is known, the photograph can be seen to reference exclusions, denial of access, the cutting up of space, and, by extension, the consequent social frictions. It becomes a picture that parodies

FIGURE 9.8 John Davis or Alfred Tattersall, "Man Seated on a Footbridge, Samoa," early 1890s

the typical paradisiacal view and that, further, implies that such visions of tropical Samoa obscured or diverted attention from the demonstrably actual or real in a sociopolitical sense.

Similarly, figure 9.9 raises questions about the complexities of contested occupation of space. This photograph taken in a Samoan village, probably by Davis, has been variously titled, according to its context of use. For instance, in Augustin Kramer's scientific *The Samoa Islands* it is "A Modern Tomb of a Chief," while in the photo archive of the Alexander Turnbull Library in Wellington it is "Samoan man in traditional costume, standing in front of a wall. A woman may be seen, partly obscured, behind the wall at right."[38] Kramer's title points to his prime concern with building an inventory of Samoan material culture for the anthropological record. The Turnbull Library's description blandly focuses on what the photograph is immediately seen to denote. Both titles, or descriptions, overlook what was going on in the photograph and, specifically, the encounter between its figures and spaces, the "inside," with the photographer and viewer "outside." For the spatial organization and orchestration of the scene presupposes the presence and indeed participation of another party in a space continuous with the figures in the photograph.

The man is represented close to the front plane of the image, as if almost in the same space as the photographer and viewers—a quality accentuated by the mounting of another photograph cutting him off at the

FIGURE 9.9 John Davis, "Samoan Man Standing before a Tomb," early 1890s
(Alexander Turnbull Library, National Library of New Zealand, Wellington)

knees.[39] His forcefully upright stance, his musculature emphasized by his
arms tensely crossed over his torso (the epitome of the aggressive/defen-
sive body language), constitutes a taut, ready-to-spring configuration,
sharpened by the look in his face. It is a hard, forbidding look, the look of
someone whose space is not to be crossed. The close view accentuates the
eye language, too. He is eyeing the photographer, and viewer, sizing them
up, as if about to say, "Keep out. No entry." Access to the space beyond
is blocked by him—to both the village to the left and, more specifically,
the *fale* or house to the right. That there is no access is reinforced by the
stone structure immediately behind him. As a tomb, it connotes the sacro-
sanct, that which is not to be entered. As a compositional element, it functions as a barrier, largely closing off any spatial recession and impeding any
open view of what is beyond, only fragments of which can be glimpsed.
Part of the roof of a house can be seen but not its open lower structure—
its interior, domestic, private space. And to the left there is only an obscured area in front of another partially visible building, for the depth of
field is fairly shallow, thus concentrating attention on the foreground
figures and event.

Equally loaded in this play on entry frustrated, of resistance to control
by European eyes, is that glimpse of a female figure's face, peeping over
the top edge of the tomb. The young female is mostly kept out of sight
and out of reach—in an image world in which otherwise there was a

plethora of young Samoan women displayed full-bodied, their looks usually implying welcome or availability, ease of access or entry. Insofar as images of young non-European women in colonized places have frequently been read as standing for the "body" of the desired land, possessed or about to be possessed, this photograph could work as a refutation of that typing, or at least as one that radically complicates a commonplace tropical South Seas vision.[40] However, in seeing self-possession and resistance in the Samoan subjects represented in this and other photographs, it is important not to overlook the brute realities of colonial power and exploitation. Certainly, many Samoans were aware of how photographs could work, of how the space of the photograph could be used. And this photograph can be seen as communicating a suspicion of the camera and a denial of access to Samoan spaces. Yet, despite this resistance to photographic "possession," what is photographed has been rendered the property of others, the photographer and the users of the photograph, incorporated into aspects of the colonial project. The image appeared in Kramer's anthropological text, where it might contribute to a colonial regulatory system.[41] And in the albums of the British Consul, Cusack-Smith, it could be seen as part of another kind of mapping of place and people in British interests. In these contexts of viewing, the self-possessed Samoan man was still "subjected to an unreturnable gaze," however strong his sense of independent agency otherwise was and is.[42] Overall, figure 9.9 can be seen as an image that touches, compellingly, a point of uneasy meeting between Samoan and European, making visible how problematic the business of boundaries was. There were limits to what could be entered and what could be made visible. What, and where, were the boundaries? In the Pacific, it has been observed, boundaries could be so numerously layered and complex as to defy adequate representation.[43]

It has been claimed that photography in places subject to colonial interests was integral to the operations of European imperialism; that various indigenous peoples and their lands were subject to a dominating gaze; that photographic representations were essential to the drives to map and control.[44] While there are plenty of Samoan photographs that lend themselves to such an argument, it is a fairly blunt instrument. Some key features of the operation and continuing life of photographs have been overlooked—notably, that their meanings and values are not necessarily singular and fixed, either in themselves or at any one point in time, but tend to be fluid, unstable, and contestable. And photographic representation was not necessarily or only an instrument of colonial or imperial domination or, simply, a reflection of colonialist ideology.

While the photographs discussed in this chapter might appear anomalous in the context of the larger and better-known body of Samoan im-

ages, they may not be so in relation to the complicated web of sociocultural and political relationships in colonial-period Samoa. There are, in
fact, many similar photographs in the archives, even if they are infrequently reproduced or exhibited. What these photographs bring into vis-
ibility is a sense of the simultaneity of diverse and conflicting views in, and
of, tropical Samoa; and they suggest that specific and local encounters
shaped differing picturings, even in the work of the one photographer.
Apia of the mid- to late nineteenth century has been characterized as a
"borderland."[45] These photographs, too, can be seen to constitute a kind
of borderland, in which relationships cannot be marked out with certainty
but are characterized, rather, by a shifting quality. In this context, tropicality emerges as a complex of intersecting phenomena and conditions, at-
titudes and experiences. The tropical, or what was frequently so regarded,
was certainly not natural, nor was it simply constituted by that set of opposing and limiting stereotypifications, as has commonly been asserted.
Gauguin's imagined tropics may have contributed to an image of the South
Pacific that is still a staple of touristic advertising and promotion. But the
tropics, as photographed by Davis, Tattersall, and others, shifts our attention markedly from that comfortable and comforting vision, questioning
it, undermining it, and offering a more compellingly human image.

Returning Fears: Tropical Disease and the Metropolis

ROD EDMOND

In this chapter I am concerned with the entanglement of the tropics and the metropolis at the end of the nineteenth century. It begins with the widespread belief that Europeans could not survive long in the tropics, moves to the fear of so-called tropical diseases invading metropolitan centers, and finishes with the contemporary view of European cities such as London as sharing some of the characteristics and effects of tropical zones. My focus in this chapter will be on disease and degeneration, with leprosy providing the particular example. The underlying argument will be the impossibility of understanding what was meant by the tropics at this time without also taking account of the national self-constructions of the period.

Geographically, the tropics are defined by those latitudes 23°27′ north and south of the equator, corresponding to the equinoctial spheres represented by the circles of Cancer and Capricorn. The region between these parallels, often referred to in the past as the "torrid zone," was of growing interest and concern from the second half of the eighteenth century as European empires spread across the globe. One indication of this was the increasing use of "tropic" as an adjective, as for example when Wordsworth wrote in his famous sonnet "September 1, 1802," on the expulsion of a black woman from France, "Yet still her eyes retained their tropic fire."[1] This usage was soon eclipsed by the adjectival form "tropical," whose applications multiplied during the nineteenth century and came to include, for example, disease and clothing as well as the more customary references to geography and climate. From around midcentury, it also took on new figurative meanings, such as ardent or luxuriant.

Whereas at the beginning of the eighteenth century the primary meaning of "tropical" would have been "of the nature of tropes," that is, "metaphorical or figurative," by the middle of the nineteenth, this had been crowded out by its ramifying climatic and cultural meanings. By the early twentieth century, these had come to require a further term, "tropicality," to describe the state or condition of being tropical.[2] As I use it in this chapter, "tropics" is emphatically a relational term, requiring the antonym "temperate" for its full significance.

During the nineteenth century, the relation between the world's temperate and tropical zones was increasingly understood as difficult, even antagonistic. Culturally, the tropics became Europe's other, a thick belt around the middle of the globe that white men entered at their peril. This had not always been the case. A hundred years earlier, although the distinctiveness of the tropical environment and its maladies was acknowledged, Europeans living in the tropics were believed to be capable of acclimatization through residence, a process commonly referred to as "seasoning."[3] This had been described by Benjamin Moseley in *A Treatise on Tropical Diseases* (1787). Moseley argued that although European animals degenerated in the West Indies he had not observed that humans did.[4] European newcomers could expect some initial disorder, but "seasoning" would "generally have a speedy termination."[5] Indeed, Europeans could benefit from exposure to the tropics: "Men generated from the coarser materials of northern melancholic matter, who on their native soil were intended to vegetate, labour, and die, often acquire an expansion of soul, removed to warmer climes. They ripen in the sun. They get ideas in spite of nature."[6] James Lind, in *An Essay on Diseases Incidental to Europeans in Hot Climates* (1768) was slightly more cautious, drawing a comparison between men and plants removed into a foreign soil and emphasizing that "the utmost care and attention are required, to keep them in health, and to insure them to their new situation."[7] Lind was in no doubt, however, that seasoning would occur, even if it might require assistance. In like manner, his *Treatise on the Scurvy* (1753) aided the health of seamen undertaking long voyages. The problem of acclimatizing to the tropics was soluble; inexperience and ignorance were at least as much to blame as "the malignant disposition of the most unwholesome climate."[8]

This optimistic view of European adaptability slowly faded, even as the actual mortality rates of Europeans living in the tropics declined.[9] By the time of James Johnson and James Ranald Martin's *The Influence of Tropical Climates on European Constitutions* (1841), the idea that men adapted better than animals to a radically different climate was being reversed: "While animals adapt over several generations to changed circumstances, that successors of those who move to tropical climates *gradually degenerate* . . . cannot easily be disproved." [10] There were a number of reasons for this. Climatic theories of racial difference were being reinforced by more essentializing biological ones. The expansion of European empires, with increasing contact between Westerners and indigenes in colonial settings, and between colony and metropolis, also contributed to a sharpening sense of racial difference closely associated with distinct climatic zones. And in the last quarter of the nineteenth century, fears arising from these increasing contacts were intensified by the emergence of the germ theory of disease and the knowledge that invisible microorganisms could unwittingly be brought back to Europe from tropical climes, even by apparently healthy human carriers. This growing belief that the tropics were dangerous fever nests, probably uninhabitable by Europeans for any prolonged period, greatly complicated the fact of their increasing economic and imperial importance for Europe and the United States. Imperial powers had a problem.

A number of influential works at the end of the nineteenth century addressed the dilemma. One of these was Charles Pearson's *National Life and Character* (1893). Pearson had been professor of history at King's College London before emigrating to Australia where he became minister of public instruction in the Victoria State Parliament and was a profound influence on the generation of politicians who oversaw the introduction of federation in 1901.[11] In *National Life and Character,* Pearson argued that because the population of tropical zones increases more rapidly than that of temperate ones, the "higher races" will become hemmed in, prevented by constitutional incapacity from living in tropical zones, and threatened at home as the inhabitants of the tropics swarm across the world. Australia offered a telling example. On the one hand, it seemed to provide a solution to the increasing lack of space in temperate zones: "The natives have died out as we approached; there have been no complications with foreign powers; and the climate of the South is magnificent." But although Australia did not have a native problem, Aboriginal people having obligingly anticipated their inevitable fate, it was acquiring an immigrant one: "We know that coloured and white labour cannot exist side by side; we are well aware that China can swamp us with a single year's surplus of population; and we know that if national existence is sacrificed to the working of a few mines and sugar plantations it is not the Englishman in Australia alone, but the whole civilised world, that will be the losers." [12]

When the Australian prime minister Edmund Barton introduced the Immigration Restriction Bill to the recently constituted Federal Parliament in 1901, he read several passages from *National Life and Character.*[13] Pearson had argued it was unlikely that "the white race" could acclimatize to the northern parts of Australia, those closest to tropical zones.[14]

The country, therefore, was not so much a last chance for the white races as a vivid example of an inexorable process: "The black and yellow belt, which always encircles the globe between the Tropics, will extend its area, and deepen its colour with time."[15] Imperial expansion, in fact, contained the seed of its own destruction. By half-civilizing native peoples and introducing them to European technology, "the races that are now our subjects . . . will one day be our rivals. . . . We shall wake to find ourselves elbowed and hustled . . . thrust aside by peoples whom we looked down upon as servile."[16] The entanglement of the temperate world with the tropical threatens the end of those races "that have taken their faith from Palestine, their laws of beauty from Greece, and their civil law from Rome."[17] Pearson's imperial version of involution theory, with the worst driving out the best, has strong parallels with the contemporary fear in Britain that a degraded urban working class was contaminating metropolitan society from within. Common to both was a fear of rampant breeding, with a differential birth rate working to the detriment of civilized society at home and in the empire.

Pearson's book was widely discussed in Britain and the United States. Gladstone was fascinated by it, and Theodore Roosevelt wrote to its author of the impact it was having in Washington.[18] Roosevelt also reviewed *National Life and Character* in the *Sewanee Review*. Although finding it "unduly pessimistic," he was struck by Pearson's argument about the declining white birthrate and "the teeming population of China": "As the world is now, with huge waste places still to fill up, and with much of the competition between the races reducing itself to the warfare of the cradle, no race has any chance to win a great place unless it consists of good breeders as well as good fighters."[19] Marilyn Lake suggests that Pearson helped inspire Roosevelt's new imperial project, with "race suicide" now becoming one of the American's favorite themes.[20] And Henry Adams, after a visit to Cuba and the Bahamas in 1894, wrote to a friend: "I am satisfied that Pearson is right. . . . In another fifty years . . . the white races will have to reconquer the tropics by war or nomadic invasion, or be shut up, north of the fortieth parallel."[21]

Another influential contribution to this late nineteenth-century two-world theory was Benjamin Kidd's *The Control of the Tropics* (1898). In Kidd's previous book, *Social Evolution* (1894), which was widely read and translated, he had argued that the subjugation, even eradication, of inferior races by the Anglo-Saxon peoples was inevitable. *The Control of the Tropics* originally appeared as a series of articles in the *Times* in 1897 and again was extensively noticed.[22] For Kidd, the "great rivalry of the past" had been won by "the English-speaking peoples," but the "great rivalry of the future is already upon us. It is for the inheritance of the tropics."[23]

Kidd focused on the economic significance of the tropics for Britain and the United States, whose combined trade with these regions was 44 percent of their total trade with the rest of the world.[24] However, political and administrative relations with these backward regions on which modern life so heavily depended "are," he said, "either indefinite or entirely casual."[25] This is an urgent matter because Kidd, like Pearson, did not think the tropics could be settled by Europeans: "In the tropics the white man lives and works only as a diver lives and works under water."[26] This inability to acclimatize was, for Kidd, a result of evolution. The human race had developed northward from the tropics. The "natural inhabitants" of the tropics had remained like children and were incapable of developing the tropical world themselves. He cites the example of the West Indies, which since the end of slavery had become like a former civilization: "Decaying harbours . . . stately buildings falling to ruin . . . deserted mines and advancing forests."[27] Here, Kidd is different from Pearson. The ascendancy of Western peoples and civilization is beyond doubt because it is underwritten by evolution, and he questions Pearson's view of Europeans being elbowed out of the way.[28] Rather than a vigorous and expanding tropical world, Kidd fears a sedentary and regressive one. Nevertheless, both see the relation between temperate and tropical worlds as the most pressing question of their time, and both agree there is an acute problem of managing the tropics because of the European inability to live there.

A variant on the evolutionary theories that underlay most turn-of-the-century discussion of the relation between Europe and the tropics is found in Charles Woodruff's *The Effects of Tropical Light on White Men* (1905). Woodruff had been a U.S. army doctor in the Philippines. He rejects the idea that the human race had evolved in the tropics: "That process required a cold severe environment which killed off all except the most intelligent in every generation . . . and thus caused an evolution of the large human brain." This drives an even sharper wedge between temperate and tropical zones; tropical peoples are permanently fixed at an inferior level of development. By the same token, however, they inhabit a world the white man is unable to share: "A species is sharply limited in its northern and southern extensions and though it may be found over longer distances east and west it is never found out of its zone. Migration would be followed by extinction sooner or later for acclimatisation is not possible."[29] In Woodruff's case, the tropics became a distinct "zoological zone" inhabited by an equally distinct "anthropoid type."[30] When any experienced change of climate is extreme, as for example in the case of white men in India, reproduction ceases altogether; there are no third-generation Europeans in India. A version of this is found in Rudyard

Kipling's *Kim* (1901), when the narrator remarks of Kim's initially rapid progress at his school: "At St. Xavier's they knew the first rush of minds developed by sun and surroundings, as they knew the half-collapse that sets in at twenty-two or twenty-three." [31] Prolonged exposure to tropical light caused the European to "break down," a recurring phrase in Woodruff that comes to mean something almost as literal as physical decomposition. Where the change of climate is less abrupt, as with the Negroes in the United States, survival is longer but extinction nevertheless inevitable. [32]

Woodruff, like Kidd, is less troubled than Pearson at the prospect of being overrun. All of them, however, are agreed that the tropics are troublingly other; difficult to manage and difficult to manage without. Imperial expansion, and the increased movement of trade, peoples, and germs that resulted, meant that the relation between temperate and tropical worlds needed to be understood in terms that took account of the greater activity between these two zones. At the same time, however, it was becoming axiomatic that Europeans could not settle and reproduce in tropical countries, while an expanding tropical population was threatening to burst out of its zone and into the temperate civilized world. Pearson and Kidd are both obsessed with the idea that the temperate regions were "filling up" and running out of space—hence the language of elbowing, hustling, swarming, and swamping. By the turn of the century, a historically particular view of the tropics as inalienably other had engrossed metropolitan cultures.

Jack London's story "The Unparalleled Invasion" (1910) provided a fascinating though rebarbative response to these concerns. Imaginative writers have freedoms that are denied to social and imperial theorists, and London's story employs counterattack as the way out of the impasse reached by Pearson and Kidd. Projected forward into the 1970s, it describes the irresistible expansion of China during the twentieth century: the alarming "fecundity of her loins" has resulted in there being "two Chinese for every white-skinned human in the world. . . . There was no way to dam up the over-spilling monstrous flood of life." Faced with this overwhelming threat, the European powers and the United States agree on a "Great Truce" and decide on counterattack (Pearson too had stressed the need for "temperate unity"). Attack takes the form of biological invasion by airship, with millions of glass tubes containing deadly microbes rained down on China. Western armies and navies are massed at the frontiers to prevent escape, an offensive blockade cutting off emigration at the source. China "perishes," and the Western powers solemnly agree never to use bacterial warfare against each other. In the case of China, though, it has been a justified response to the swarming of a race that threatens to contaminate civilization. [33]

London's story is a projection of the disease fears produced by the large-scale movements of population between metropolis and colony, and the discovery of the role of microorganisms in the etiology of disease. It responds to the imagined flow of deadly germs from East to West by reversing their direction, ensuring that Europe gets its revenge in first. Germ theory is often said to have undermined geographical and climatic explanations of disease, and to some extent this is true. It certainly focused attention more on the native subject as disease carrier than on the tropics as a pathological site. But native peoples continued to be thought of in terms of the tropical places they inhabited, and these places in turn reinforced the belief that European bodies were very different from those of their colonial subjects. Germs, those invisible carriers of disease, intensified the view of both tropical places and peoples as toxic. And one important consequence of this was a new, bacterially derived way of stigmatizing immigrants as the bearers of germs, in place of earlier forms of stigmatization that had seen immigrants as producing unsanitary and disease-making conditions.[34]

The late nineteenth-century construct of tropical medicine was one of the prime means by which this increasingly pathologized relation between tropical and temperate zones was to be described and regulated. Historians nowadays frequently, and more accurately, refer to this turn-of-the-century branch of medicine as imperial or colonial medicine.[35] Tropical medicine as a category ignored diseases like measles and influenza that made terrible inroads into indigenous populations and concentrated instead on diseases believed to be specific to tropical climes and to which Europeans were especially susceptible. In effect, it was a metropolitan imposition on the equatorial regions of the world at a time when the relation between temperate and tropical zones had become particularly troubling. Germ theory meant that the disease maps of medical geography were no longer adequate. A new kind of mapping was needed that would take account of the ubiquity of germs without relinquishing the temperate/tropical divide. A germ- and vector-based medicine in which the tropics, and in particular tropical bodies, were pathologized in modern scientific terms as the natural site of deadly microbes was well-adapted to this need.[36]

The founding text of tropical medicine, Patrick Manson's *Tropical Diseases* (1898), conceded that the category was "more convenient than accurate," excluding many diseases found in the tropics while including some that were not confined to these regions.[37] Manson's classification runs into problems with leprosy, for example, which as he admits does not depend on climatic conditions or terrain.[38] Once prevalent in Europe, now mainly found in tropical zones, it had also been spreading in Norway

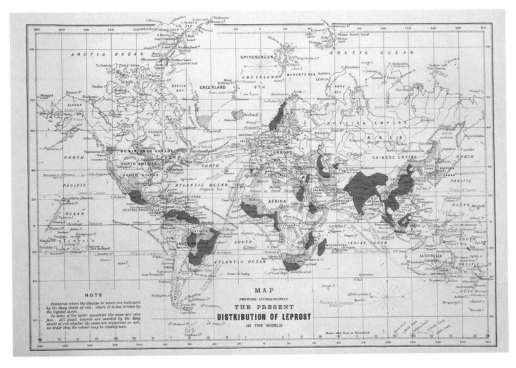

FIGURE 10.1 "Map Showing Approximately the Present Distribution of Leprosy in the World," from George Thin, *Leprosy* (London: Percival & Co., 1891) (Wellcome Library, London)

during the nineteenth century (cf. fig. 10.1). And although Manson had no doubt that it was caused by the leprosy bacillus and conveyed by lepers, no one had yet succeeded in cultivating the bacillus or transmitting it by inoculation.[39] Further, unlike the emerging type of a tropical disease with rapid onset and vivid symptoms, leprosy developed very slowly and was difficult to diagnose in its early stages. Yet Manson had no doubt that leprosy was a tropical disease, "an important element in the pathology of nearly all warm countries."[40] His account of its later stages illustrates and emphasizes the atrophy and deformation wrought by the disease, and this disfiguring was clearly an important element in its designation as "tropical." Interestingly, the illustrations of leprosy that Manson included in the early editions of *Tropical Diseases* were of Europeans. Nancy Stepan contrasts the relative restraint and dignity of one of these images with Manson's more dramatic photographs of Samoans suffering from elephantiasis of the breasts and scrotum. She argues that the photographs of elephantiasis, by focusing on the sexual organs, reveal the connections Europeans made between blackness, sexuality, pathology, and the tropics. These images of elephantiasis, Stepan concludes, offered a more vivid representation of "tropical enormity" than other tropical diseases: "Elephantiasis signed itself on the body in an obvious and dramatic fashion; it was a frightening condition, largely of people faraway, who *could* be shown naked. . . . [It] showed all that was believed to be pathological about the tropics: its environment, its racial make-up and its sexuality."[41]

This perceptive argument can, I believe, be nuanced. There was, in fact, no physical reason why leprosy, too, could not also have been used to illustrate tropical deformity. Within Judeo-Christian cultures it was, par excellence, the disease that signed itself on the body, and it had repeatedly been sexualized, indeed often confused with syphilis. The point, I feel, is not so much that natives were chosen to illustrate elephantiasis (they were also used, for example, to illustrate sleeping sickness in the early editions of *Tropical Diseases*) but that Europeans provided the subjects for leprosy illustrations. There was nothing inherently "tropical" about leprosy at all. On the contrary, there was long experience of the disease within Europe and a complex tradition of regarding it with both horror and compassion. The leper was "unclean" and reviled but also a special object of Christ's compassion. Although, as Stepan points out, elephantiasis was no respecter of race, Europeans were only at risk from it in tropical countries.[42] Leprosy, by contrast, was as much a European as a tropical disease and was feared to be returning to metropolitan centers. Manson makes specific reference to this, in the widely cited case of a man from Dublin, who had never journeyed further from home than England, contracting leprosy from his brother who had lived in the West Indies.[43] The choice of a European to illustrate the effects of the disease can, I think, be taken as a recognition of its highly ambiguous status as a "tropical disease" (see fig. 10.2).[44] Unlike elephantiasis, leprosy was not only faraway but also close at hand. Just how close was beginning to trouble the authorities in several European capitals.

The head and shoulders illustration of leprosy that Manson used in his first three editions suggests a further refinement of this argument. Although the figure is clearly European, the symptoms of a developed case of tuberous leprosy result in some blurring of racial difference.[45] In particular, the elongation of the face and the thickening of the features produce an effect of racial ambiguity. The very distinctions that tropical medicine was intended to affirm become equivocal in the case of this geographically unstable and shape-shifting disease. Manson, nevertheless, binds leprosy into the category of tropical medicine. Like many so-called tropical diseases it was actually a disease of poverty, malnutrition, and unsanitary conditions, but the category allowed it to be associated with natural rather than social and economic factors.[46] So for all the ambiguity of

FIGURE 10.2 "Patient with Nodular Leprosy of Six Years Duration," reproduced in Patrick Manson, *Tropical Diseases* (London: Cassell & Co., 1898) (Wellcome Library, London)

its status, Manson included leprosy as one of numerous diseases to which tropical natives are inherently susceptible.

The new science of tropical medicine, therefore, can be understood as an attempt to put a fence around Europe and around the European in the tropics. If the temperate and tropical zones were really two distinct worlds then it was important to preserve or protect the former from the latter. The discovery that invisible germs spread by human contact could infect bodies and spread illness threatened this division, as did the increasing flow of people and goods between these zones. From this point of view, tropical medicine becomes an example of what Laura Otis has termed a membrane, an attempt to preserve identity by sealing the metropolis and the colonizer from the world they were colonizing.[47] Such membranes, however, were inevitably semipermeable, mediating between two worlds as well as fixing the difference between them, an imperfect filter rather than a barrier. This, in turn, threatened the boundaries that such membranes were intended to define. Leprosy—itself in so many ways a boundary disease and sitting uncomfortably within the category that was intended to describe and contain it—provided a vivid example of this fear of transgression.

Official concern about the spread of leprosy in the British Empire dates from the 1860s. Reports from the West Indies prompted the Colonial

Office to ask the Royal College of Physicians to conduct an empire-wide survey, which concluded that leprosy was hereditary not infectious, that it was almost exclusively a native disease, and that it posed no real danger to imperial well-being.[48] These findings were counterintuitive, and although the hereditary theory of causation remained the official government position until the 1890s, it was never widely accepted. By the 1880s alarm was spreading at the apparent rise in the incidence of the disease around the empire and fears were beginning that it would return to Europe in a kind of bacterial countercolonization, reinfecting the metropolitan centers from which it had long disappeared. In 1889 Father Damien, the Belgian priest famous worldwide for his work at the leper colony on the island of Molokai in the Hawai'ian Islands, died of the disease. This marked the virtual end of the idea that leprosy was hereditary and that Europeans were more or less immune.

In the same year as Damien's death, Henry Press Wright published *Leprosy: An Imperial Danger,* an alarmist warning about the dangers of a free trade in disease in which leprosy would return to Britain from tropical colonies. Also in 1889, a leading British doctor, Sir Morrell Mackenzie, published a similar warning in a widely noticed essay in the *Nineteenth Century:* "Leprosy has before now overrun Europe and invaded England, without respecting the "silver streak" that keeps off other enemies; and it is perfectly conceivable . . . it might do so again."[49] I am unsure whether the "silver streak" refers to the English Channel, which John of Gaunt saw as the moat protecting "this scept'red isle . . . This fortress built by Nature for herself / Against infection," or to mercury (quicksilver), a common treatment in the past for many ailments, especially venereal diseases.[50] Whether Mackenzie was speaking of the return of leprosy or the inability to treat it, there is no doubt of the alarm it was causing.

Isolated cases of leprosy in Britain were being reported. One such case was discovered in a Scottish boarding school in 1887. The infected pupil was British but had grown up in the West Indies and was said to have been infected as a result of arm-to-arm smallpox vaccination of a native child from a leprous family. The boy was asked to leave the school, which then faced a legal action for breach of contract. The judge found in favor of the school, ruling that it was "absolutely necessary to take action so as to prevent the whole of them becoming lepers by contagion. . . . The existence of such a disease in the midst of a community of boys . . . was calculated to create such terror as to impair the usefulness of the institution."[51] The school concerned was Fettes College, Tony Blair's alma mater.

In the weeks following Damien's death in 1889, the London papers were full of reports about leprosy. Two lepers were exhibited at a meeting of the Epidemiological Society, an Indian army doctor warned that leprosy would return to Britain unless segregation policies were adopted in the empire, and alarm was expressed at the prospect of leprosy stalking the streets of London.[52] The *Pall Mall Gazette* commented on the meeting of the Epidemiological Society: "One of the lepers . . . was an old man who did not show any outward sign of the disease, and who might have been passed in the street without any suspicion that he was a confirmed leper."[53] Germ theory suggested that we lived surrounded by invisible enemies, and the German bacteriologist Koch had formulated the idea of the healthy carrier, something that rendered everyone potentially suspect.[54] Hitherto, leprosy had at least been thought of as a highly visible disease, its horrifying physical symptoms a warning to others. Now it seemed not.

A few days later, at the inaugural meeting of the Father Damien Memorial Committee, the Prince of Wales announced there was a leper working in the London meat market. The *Times* reported that the man in question was one of the two recently exhibited at the Epidemiological Society. Eventually he was named as Edward Yoxall, sixty-four years old, who sold ox tails, sheep heads, and offal from Smithfield to the poor of Hoxton (fig. 10.3). His hands and feet had been affected "in so horrible a manner as to render it almost incredible that he should be able to do any kind of work." As a young man he had traveled the world as a sailor, and it had taken thirty years after coming ashore for his symptoms to develop.[55] This meant that if the disease had not been contracted in Britain it must have incubated for several decades, "the germ," as one doctor put it, "dormant like 'mummy' wheat!"[56] Damien's death had ended any lingering belief in European immunity. This case suggested there was no geographical or climatic immunity either. It also added to accumulating evidence of the inexplicable caprice of the disease.

The story grew and faded, as health-scare stories do. Edward Yoxall was secured in Whitechapel Infirmary, his wife received assistance from the poor box, and the Prince of Wales sent a check.[57] Similar stories came and went, usually prompted by some other event that brought leprosy to public attention. For example, the first international conference on leprosy in Berlin in 1897 was followed by another "startling discovery" of a case of leprosy in London.[58] The specter of leprosy in the streets of London continued to haunt from time to time, reviving a long-running debate over the question of dedicated provision for lepers in Britain. The argument in favor was met by the often-voiced fear that a leper asylum in Britain would attract lepers from around the empire to the metropolis, thereby materializing the fears that discussion of the disease always provoked. Eventually a leper colony was established in rural Essex, at East Hanningfield, in 1914, to the consternation of local residents.

This is not the place to pursue the history of leprosy in Britain around

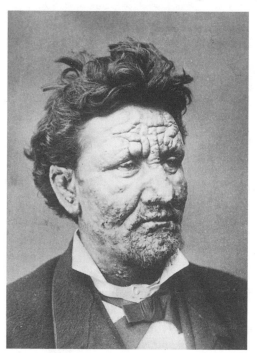

FIGURE 10.3 "Patient with Nerve Leprosy" (probably Edward Yoxall), from George Thin, *Leprosy* (London: Percival & Co., 1891), plate 2 (Wellcome Library, London)

the turn of the twentieth century. Nor am I claiming that a fear of its return was a constant factor in British social life at that time. I am arguing, however, that leprosy provided a particular focus for concern about the tropics, for the disease fears prompted by the mixing of racialized bodies, and that this directly affected the metropolis itself as well as the tropical colonial world. In the closing pages of this chapter, I shall examine some of the ways in which leprosy also provided a powerful metaphor, and an extensive repertoire of associations, for a wider concern with metropolitan degeneration at the turn of the century.

By the 1880s a discourse of the tropics was increasingly being applied to urban slums. Although the comparison of Britain's cities to a tropical jungle was hardly new—Dickens, for example, had made it in his description of Coketown in *Hard Times*—this trope became extensively used and elaborated in the final decades of the century.[59] If any one work fixed this identification in the public mind it was William Booth's *In Darkest England and the Way Out* (1890). The deliberate echo of Henry Morton Stanley's *In Darkest Africa,* published in the same year, and the extended analogy between England and Africa of Booth's opening chapter, was arresting; within a month it had sold 115,000 copies. The image of "darkest England" provided a telling environmental metaphor with which to express a concatenation of fears about the health of the metropolis and the empire.[60] *In Darkest England* opens with an extended description of the immensity and near-impenetrability of the African forest, which is contrasted with the physically stunted and barely human pygmies who inhabit this world. This overwhelming and implacable environment that renders helpless a stunted and malformed people is then pathologized in terms that allow Booth to pin the comparison between tropical and so-called civilized worlds: "Darkest England, like Darkest Africa, reeks with malaria. The foul and fetid breath of our slums is almost as poisonous as that of the African swamp. Fever is almost as chronic there as on the Equator. Every year thousands of children are killed off by what is called defects of our sanitary system. They are in reality starved and poisoned."[61]

Many of Booth's critics pointed out that *In Darkest England* was less original than it claimed, and his use of the city/jungle analogy was no exception. The comparison of London to the tropics, with disease or ill-health a particular focus, can be found in a range of different kinds of writing at this time. George Gissing's novel *The Nether World* (1889), for example, represents London slum life as a never-ending struggle for survival in which the most animalistic prevail. Exemplifying this is the character of Clementina Peckover, whose "forehead was low and of great width," whose nose "had large sensual apertures," and who is likened to a savage running wild. This relative commonplace becomes more interesting when the apparent contrast between her physical health and her moral degeneration is shown to be deceptive: "Her health was probably less sound than it seemed to be; one would have compared her, not to some piece of exuberant normal vegetation, but rather to a rank, evilly-fostered growth. The putrid soil of that nether world yields other forms besides the obviously blighted and sapless."[62] Urban degeneration, in other words, took a variety of physical forms and was not always immediately apparent. The fittest were not even necessarily fit.

More commonly, however, it was the blighted and sapless forms of human life that preoccupied fin de siècle observers of the metropolis. James Cantlie's *Degeneration amongst Londoners* (1885) compares Londoners to those returning from the tropics, "blanched and pale . . . suffering from Asiatic diarrhoea or dysentery."[63] Those with parents also born in London grow up physically underdeveloped, of scrofulous aspect, with squints and misshapen jaws. Suffering from "urbomorbus," or "city disease," their bones fail to develop, becoming soft and spongy and unable to support the body. The only comfort is that these people cannot reproduce: "Nature steps in and denies the continuance of such."[64] In this they are like third-generation Anglo-Indians for whom the "attaining of adult years is impossible."[65] The colonial tropical problem of "noncontinuance" has now come home to the metropolis. Significant sections of the British population are no longer even acclimatized to their own country. Cantlie concludes: "It is beyond prophecy to guess even what the rising degeneration will grow into, what this Empire will become."[66]

Cantlie was a doctor whose interest in the physical condition of the urban poor had developed while working at London's Charing Cross Hospital. He had then served in Egypt where he experienced a cholera epidemic and became interested in so-called tropical diseases. In 1887 he took over Patrick Manson's practice in Hong Kong where he worked for a decade before returning home and assisting Manson to establish the London School of Tropical Medicine in 1899. An original member of the staff of that school, and founder editor of the *Journal of Tropical Medicine and Hygiene,* he had a distinguished career in tropical medicine and was subsequently knighted.[67]

Cantlie's writing exemplifies the close relation between fears of disease and degeneration in nation and empire. In 1890 he published *Leprosy in Hong Kong,* which vividly captures the contemporary fear of the migrant leper. As Cantlie describes it, the free port of Hong Kong was being overrun by Chinese lepers escaping from "the wretched leper villages" of the mainland."[68] In 1897 the British National Leprosy Fund published Cantlie's *Report on the Conditions under Which Leprosy Occurs in China, Indo-China, Malaya, the Archipelago and Oceania,* which opens with a parallel between the end of the nineteenth century and "the dim centuries" of the past (not unlike Booth's opening comparison between Britain and Africa); once again, "civilised and uncivilised man confront each other with the canker of incurable leprosy in their midst."[69] Cantlie argued the need for the medical inspection of Chinese migrants using European shipping lines and for the deportation rather than segregation of all migrant Chinese lepers; segregation, he argued, merely produced further centers of infection. Cantlie concludes with a picture of leprous Chinese spreading out across the Pacific and "tainting the world."[70] Back in Britain at the time of the Anglo-Boer War when there was concern and a good deal of alarmist writing about the physical deterioration of the nation, Cantlie returned to the degeneration theme with the publication of *Physical Efficiency* (1906). Once more he diagnosed a single problem in which empire and metropolis were intertwined—that of "the wastage of life . . . in the cities of our own country . . . [and] the loss of health and life attaching to the governing and commercial development of our Crown colonies, of the great Empire of India, and of many countries lying within the tropics."[71]

This linkage of degeneration at home and in the empire was particularly influential at the turn of the century. In 1885 Cantlie's *Degeneration amongst Londoners* had been widely discussed but also ridiculed, more because of its curious theory that outer London was depriving the inner city of ozone by "pre-breathing" its air than because of its account of physical deterioration.[72] In the foreword to *Physical Efficiency,* however, Sir James Crichton-Browne, FRS, presents Cantlie as a prophet once scorned but now heeded.[73] Cantlie's influence is apparent, for example, in Jack London's *The People of the Abyss* (1903), an account of East End life based on his brief residence among the poor of the metropolis. London describes the "vigorous strong life" of rural England pouring into the city and "perish[ing] by the third generation."[74] Somewhat contradictorily he also sees this condemned East End spreading outward, invading and swamping middle-class suburbs that have hitherto enjoyed "bits of flower garden, and elbow room, and breathing space, replacing this airy comfort with "the black night of London."[75] This double movement between city and country parallels the degeneration of Europeans in tropical colonies and the tropical overspill that was believed to be threatening the temperate world. London pictures the East End as a jungle inhabited by gorillas and wolves, "a new species, a breed of city savages."[76] With the metropolis "drained . . . of its life blood," Britain's "vast empire is foundering."[77] London, like Cantlie, associates generational decline, bad air and "disease germs" with tropical life at home and abroad.[78] Physical decay is always central to this mode of thought.

The gradual but inexorable destruction of the body caused by leprosy offered a vivid way of figuring individual and social degeneration, providing an image as well as a disease with which to fix a culture's fears. Oscar Wilde's *The Picture of Dorian Gray* (1891) is a notable fictional treatment of the degeneration theme that draws on the figure of leprosy in dramatizing the relation between body and portrait, itself a complex variation on the self/other binary. Dorian Gray's decaying portrait becomes a caricature of the body's slow corruption as it ages, miming the leper's condition of death-in-life. The self represented by the decomposing portrait is repeatedly associated with those diseased parts of outcast London that were felt to threaten the health and well-being of the metropolis; for example, the opium den of Malays, half-castes, and syphilitic women of the novel's night-town sequence in chapter 16. This apparently all-too-familiar orientalizing strategy, however, also works to undermine the aestheticizing of such cultural references elsewhere in the novel—the "dainty Delhi muslins," "Dacca gauzes," "elaborate yellow Chinese hangings," and so on in chapter 11, with which Dorian Gray attempts to embroider his West End life.[79] This is one particular example of the larger theme of the work, the impossibility of keeping degeneration at bay. The two worlds of the novel are shown to be one as its polarities of beauty and horror, health and disease, West and East, collapse together in the climactic scene of Dorian's death, his body found lying on the floor, "withered, wrinkled and loathsome of visage."[80] The boundaries between these two worlds are not so much porous as nonexistent. Degeneration is not confined to the East End but lies at the heart of metropolitan culture and "noncontinuance" extends beyond the tropics and the East End to the cultured world of upper-class London, marked by suicide and unable to reproduce itself. "Fin de siècle" murmurs Lord Henry, only to be trumped by Lady Narborough's "Fin du globe."[81]

A Sherlock Holmes story, "The Blanched Soldier," provides a further point of reference. Holmes, like Dorian Gray, often disguises himself in order to penetrate an orientalized East End, although less for private pleasures than to rescue the world from its enemies, many of which are associated with the tropical outposts of empire. The first work in which Holmes appeared, *A Study in Scarlet* (1887), had opened with Watson returning to London—"that great cesspool into which all the loungers and idlers of the Empire are irresistibly drained"—after serving in the second Afghan war. He has been shot in the shoulder, rescued from the "murderous Ghazis," and fallen victim to enteric fever, "that curse of our Indian possessions." Debilitated, enervated, "leading a comfortless, meaningless existence" in the "great wilderness of London," he is rescued from ennui by his introduction to Holmes, with whom he will patrol the city.[82] In "The Blanched Soldier," however, Holmes is on his other favorite stamping ground, the country house, investigating the strange incarceration of the son of the house since his return from the Boer War. He learns that the young man is believed to have contracted leprosy when, having been wounded in battle, he crawled inadvertently into a leper colony and spent the night bleeding in a leper's empty bed. With the help of a distinguished dermatologist, Holmes is able to reassure the family that their son merely has "pseudo-leprosy . . . unsightly, obstinate, but possibly curable, and certainly non-infective."[83] For all the comfort of its resolution, however, the story dramatizes imperial anxiety about leprosy and the fear of its return. Heightened by uncertainty about its transmission, "The Blanched Soldier" expresses and contains the fear of what would happen if this disease were to get beyond the country house and into the teeming modern imperial city.[84]

This chapter has taken a relatively narrow path through the degenerationist debate of the end of the nineteenth century. The broader debate has been well discussed by, among others, Felix Driver, Gareth Stedman Jones, Andrew Lees, and Daniel Pick.[85] My intention has been to supplement this work by drawing attention to the ubiquity of disease fears within metropolitan degenerationist discourse and the complex entanglement of these fears with the problem of health and survival in Britain's tropical colonies. A standard and still very useful work such as Gareth Stedman

Jones's *Outcast London* (1971), for example, has little to say about health and disease as such. And insofar as it considers the imperial dimension of degeneration at home, it does so entirely in terms of the arguments of social imperialists such as Arnold White and Lord Brabazon, who advocated state-aided colonization to clear London of its "redundant" population.[86] Whereas the social imperialists wished to solve metropolitan problems by removing their supposed cause to the colonies, the writers I have examined were more concerned about the threat of importing disease from the colonies. Daniel Pick's groundbreaking study *Faces of Degeneration* (1989) consciously distances itself from those who explain degeneracy as fundamentally a concern with the non-European world. Pick criticizes Fabian and Said for ignoring the reality of the fear of degeneration in Europe at this time and chooses to treat the problem solely as a national and internal one.[87] However, there is no reason to reduce this to an either/or question. To argue that the colonial and metropolitan dimensions of this concern were intimately related is not to deny the specificity of either. Colonial and metropolitan perspectives were never mutually exclusive, and the disease fears that I have been discussing are a particular example of a broader concern with the relation between metropolis and colony at the end of the nineteenth century, in which disease was both a worry in itself and a way of figuring other kinds of anxiety.

It was never the case that the fin de siècle metropolis was understood solely in terms of degeneration and disease. Andrew Lees makes clear that ideas of the city as a cultural stimulus, a site of community and a positive image of modernity were also current, while at the same time noting these were less common in Britain than in other urbanizing countries of Europe.[88] D. H. Lawrence well understood the English distaste for cities, and Raymond Williams has analyzed this structure of feeling across several centuries of national literature.[89] Even the example of a positive English view of urban life that Lees provides is within the terms of the discourse I have been examining and merely reverses its conclusion. Writing in response to Cantlie's arguments, Everard Digby insisted that "the disappearance of the city from modern life would be neither healthy nor satisfactory. . . . The city is not a disease, but a necessary organic centre of civilised life. . . . The active energetic countryman enters the town and his descendants burn away within two generations, supplying in the meanwhile all the physical and mental energy which keeps the interminable wheels of civilization a-spinning."[90] In other words, Digby accepts the "noncontinuance" argument but sees this as a necessary and acceptable cost of modernity.

Nor should it be supposed that there was complete consensus about the inability of Europeans to acclimatize to the tropics. Luigi Sambon put this idea under critical examination in the *British Medical Journal* in 1897 and won approving support from a leading article in the same issue. Sambon addresses the same problem that we saw being considered by Pearson, Kidd, and others at the start of this chapter. On the one hand, European states regard "the development of the Dark Continent as the means of relief of the overcrowding of their populations, and of securing new markets for the produce of their industries." On the other, there is "almost universal agreement . . . that complete acclimatisation of Europeans in the tropics is impossible." He refers to the belief that "the strongest blood cannot endure continuous city life for more than three generations" and cites Boudin's research that had failed to discover any "pure Parisians" who disproved this three-generation rule. However, Sambon contends that the difficulties of colonization are due to parasitism rather than climate and that acclimatization is predominantly a matter of hygiene. This was a new twist in the application of the germ theory of disease to imperialism. Previously we saw the consternation caused by the possibility of invisible germs spreading back from Europe's colonies to its metropolitan centers. Sambon, however, is using germ theory to question the nonacclimatization argument. If disease is caused by parasites or germs, then acclimatization is a question of hygiene rather than innate incapacity.[91]

The lead writer in the *British Medical Journal* spelled out Sambon's argument: "If tropical disease is not the effect of heat, but of tropical germs . . . why cannot man, by suitable hygienic arrangements, escape from the tropical diseases in the same way . . . he has contrived to escape from, or contend successfully against, so many of the diseases, equally deadly, of temperate climates?"[92] However this attempt to challenge the orthodoxy of nonacclimatization remains caught within a basic assumption of the theory it disputes. The *British Medical Journal* accepts without question that Europeans cannot labor under the tropical sun. Hygiene will not ease or facilitate that degree of acclimatization, and insofar as physiological adaptation might eventually do so, this cannot be achieved rapidly "unless . . . we hurry matters by applying to the breeding of races of men those principles of artificial selection which we apply so successfully to the breeding of domestic animals."[93] This last suggestion, it is clear, is wholly rhetorical. During the course of more than one hundred years, the debate has come full circle. Whereas Benjamin Mosley had argued that European animals degenerated in the tropics but that humans did not, the *British Medical Journal* is claiming that in spite of germ theory it is animals and not humans that can be acclimatized to work in the tropics. While it was correct that the answer to tropical disease lay in a better understanding of those recently discovered germs that caused it, a still unchallengeable assumption about the incapacity of Europeans to survive successfully in the tropics prevented germ theory from being understood as pointing to a real solution. In this way, critiques of the impossibility of acclimatization became muted variants of the pessimistic consensus they had been designed to challenge. So although there were other, more positive, images of the city than that of "outcast London," and although the theories of nonacclimatization and noncontinuance did not go unchallenged, these images and theories remained overriding. Attempts to refute them characteristically ended up closer to the point of view they had set out to challenge.

In this chapter I have sought to highlight a deepening sense of incommensurability between tropical and temperate worlds in the later part of the nineteenth century. At the same time, however, a perceived failure of European acclimatization in tropical zones was imbricated in a broader degenerationist narrative that included significant sections of the metropolitan population at home. Disease and bodily decomposition was a particular focus of these fears in tropical and metropolitan worlds, with leprosy as both a particular anxiety and a powerful way of figuring these concerns. Global imperialism had opened up a new kind of material space within the metropolis, and metropolitan constructions of the tropics were strongly influenced by contemporary fears for the health of the national body. In this way, imperial and national concerns were displaced onto each other in a process of mutual shuttling between colony and metropolis.

AFTERWORD

Tropic and Tropicality

DENIS COSGROVE

The essays collected here variously parse the discursive usages of "tropicality" as the concept evolved during the period of most intense European penetration into the geographic spaces and human worlds that bestride the equator. This period extends from the mid-eighteenth-century Pacific voyages, through Alexander von Humboldt's and Charles Darwin's South American travels, records, and writings and the European colonial grabs of Africa, Indo-China, and Oceania, to the mid-twentieth-century achievement of political independence, which laid the foundations for voices from within "the tropics" to command attention and respect beyond and for the evolution of postcolonial critical frameworks that inform the contributions to this book. But, as the term "visions" implies and as various essays here testify, Europeans did not arrive in tropical latitudes free from presumptions or anticipations about these regions of the globe. The framework of "Voyages," "Mappings," and "Sites" suggests a complex interplay of representation and experience within which encounter with actual places and peoples was mediated (if not overdetermined) through the long premodern history of "Western" ideas and images about how the world between Cancer and Capricorn might be. Negotiation of meanings has been continuous between tropical imaginings and the sensuous, embodied experiences and subsequent representations of visionaries and voyagers, merchants and missionaries, conquerors and colonists.

That negotiation did not cease with decolonization; it continues today. The intensity of imaginative geographies attached to the tropics among Europeans and peoples of the global North more generally has not diminished, as a casual survey of their appearance in leisure and tourism, product advertising, and reality television immediately attests. Together with a kind of postcolonial nostalgia, the tropes of tropicality also continue to attract interest in more "serious" art and literature: in Paul Gauguin's or Henri Rousseau's paintings, for example, and such novels as *Out of Africa, The Lover,* and *The Quiet American,* to name just three works whose film versions have achieved critical acclaim in recent years. And an intense tropicality informs not only the title but the structure, argument, and presentation of scholarly works, including some of the mid-twentieth century's most influential studies of tropical place and people: Claude Lévi-Strauss's *Tristes tropiques* or Margaret Mead's *Coming of Age in Samoa.*

In this afterword, I explore some of these continuities in order to set the foregoing essays' intense focus on what might be termed "Humboldtian" tropicalities into a broader geographical and historical context. In doing so I touch on the interplay between tropical epistemologies and what we might call the "ontological tropics." By this phrase I mean intellectual framings of the geographical spaces that lie astride the equator. My contribution traces some historical drifts between, on the one hand, the tropics as geographical spaces (whether conceived in mathematical-cartographic terms, or as a physical geography of lands, sea, and material environments, or as an ethnographic-ecological space) and, on the other hand, tropicality as a set of imagined, pictorial, and textual spaces. The three orders of tropical reality I outline here are mutually constitutive, and they are consistently informed and represented by a fundamental mapping imperative that seeks at once to describe and to bound the differences between lower and higher latitudes. The interplay between the tropics as a cosmographic inscription—a physical-geographical zone of the globe— and as a distinct ethnographic region is of course fundamental to the very concept of tropicality. Throughout this discussion it behooves us to remember what O. H. K. Spate has claimed of the Pacific—whose history and geography in Western renderings has been closely coupled with that of the tropics.[1] While the tropics, like the Pacific Ocean, are genuinely a European artifact and can thus be studied as such, this is not true for the peoples who inhabit the geographical regions contained within the tropics, who are artificers themselves. Thus, in rehearsing—even with critical intent—the ways in which Europeans so closely and outrageously have bound tropical ethnography into a mutually deterministic embrace with the physical environments of the tropics, we risk perpetuating the silencing of voices speaking from within tropical space. The authoring of this collection has not escaped that dilemma, nor indeed does my own contribution.

THE COSMOGRAPHIC TROPICS

Located between 23°27′ north and south latitude circles, the terrestrial tropics are primarily a cosmographic effect. It is worth recalling the distance of their intellectual origins from the modern world. The originating tropics are celestial rather than terrestrial markers within a geocentric cosmos—in ontological terms, idealist spaces. Ptolemy's second-century C.E. text, *The Geography,* whose reappearance in the Latin West so powerfully informed European global penetration to the west and south in the fifteenth and early sixteenth centuries, summarizes antiquity's primarily deductive global geography. It assumes a two-sphere model of the cosmos whose central, static earthly globe "has, so far as the senses can perceive, the relation of a point to the distance to the sphere of the so-called fixed stars," which are attached to the inner surface of the immense but self-contained sphere of the heavens.[2] This sphere rotates around an axis whose intersection with the celestial sphere defines cardinal north and south and whose direction of rotation determines cardinal east and west. The largest of the parallel circles traced by the stars is the celestial equator, while other parallels may be defined mathematically according to the Euclidean division of the spherical distance between poles into 180 degrees and, thus, between pole and equator into ninety degrees, counting from the latter as zero degrees.

The one great circle on the celestial sphere to interrupt the symmetrical bands of parallels is the ecliptic, which traces a line determined by the apparent movement of the sun by about one degree per day through the sphere of the fixed stars. This is the consequence of the angular divergence of some twenty-four degrees between the rotational axis of Earth and that of the solar system. In a geocentric cosmos, the sun passes between winter solstices through twelve constellations from Aquarius to Capricorn, reaching a northerly point 23°27′ above the celestial equator in Cancer and a southerly point the same angular distance below it in Capricorn. Together with the polar circles at 66°33′ north and south latitudes, the equator, tropics, and ecliptic constitute the six great circles that formed the armillary sphere, that signature of the cosmos and icon of natural philosophy into the modern age.

Tracing these astronomical circles onto the earthly globe and using them to determine five zones—two frigid, two temperate, and a single torrid zone—was formulated in Aristotle's *Meteorology* (2.5.362a32). The precise correspondence among the tropic great circles, the zones, and "climate," as the bounded regions came to be termed, was never fixed, as each of these concepts has a different origin. The circles, as we have seen, are primarily astronomical. "Temperate," "frigid," and "torrid" are empirical, though relative, terms derived from Greek ethnocentric mapping. The word "climate" derives from the *klimata* (*klima* = inclination), a series of latitudinal bands of unequal width, within which the angle of the sun and, thus, the maximum length of day supposedly does not vary significantly. Generally, seven such *klimata* were mapped by Greek geographers across each hemisphere. In simplified medieval renderings of the world picture, such as those of Macrobius or Isidore, these three terms coalesced, so that the great circles of tropics and poles came to coincide with the limits of zones of habitability characterized by distinct "climates" in the modern sense of the term (see fig. 5.1). The revival in the Latin West of Ptolemeic mapping from the early fifteenth century continued this practice, despite the fact that the great circles do not coincide with lines of latitude generally drawn at five- or ten-degree intervals. Thus on world maps, whether plotted by rhumb lines or projected with a graticule, the great circles, and especially the equator and tropics, remain the most visible markers of cosmography, the science that connects earth and heavens.

The representational impact on conceptions of global geography of the tropic lines traced on the world map is progressively more difficult to grasp fully today. Since the late 1960s, satellite photography and remote sensed imagery have revolutionized global maps and related representations. The lines of latitude and longitude have tended to disappear from contemporary world maps in favor of a more "natural"—not to say "ecological"—image, and the tropics and polar circles, if shown, have retreated to scarcely visible pecked lines. This is very different from early modern maps where the equatorial, tropic, and polar lines are represented by thick, often colored or strongly inked bars. The use of a cylindrical projection centered on the equator, for most such maps further emphasized the tropic lines while shifting the polar circles to the outer margins of the space of representation. The popularity of the double hemisphere map from about 1600 into the nineteenth century was related to a "scientific" imperative to display the most up-to-date knowledge of the world and its workings, thus the line of the ecliptic and the tropics tend to be prominently marked. Therefore, while the region lying between Cancer and Capricorn may not always have been explicitly named as the "torrid zone," as it is, for example, on Jaugeon's 1688 *Carte general contenante les mondes celeste, terrestre et civile,* it nevertheless strikes the observer as a clearly distinct central band of the globe. However, its definition, cartographically at least, remains primarily mathematical rather than environmental or ethnographic.

THE GEOGRAPHICAL TROPICS

The history of European infilling of the blank spaces between the tropic lines on its globes and planispheres is the familiar one of "geographical discovery." Paradoxically, the notion of tropicality has attached itself most powerfully to those parts of the terrestrial globe that were last to come within European purview: the Americas and the islands of the Pacific. As David Arnold shows, only in the nineteenth century was India incorporated into a tropical vision that had been framed elsewhere. Africa, the only "tropical" region within the classical *oikoumene,* was figured as a site of slavery, superstition, and disease for much of the eighteenth and nineteenth centuries, though its place in the modern discourse of tropicality emerged more fully in the course of the scramble for imperial control, when Africa became strongly connected to tropical medicine and science. This pattern is perhaps explained by the shock of discovery that attached itself to the western tropical "new" worlds of America and Oceania. Africa, India, and even Indo-China and the eastern "Spice Islands," however exoticized, were continuously part of the Western and Islamic episteme from classical times. This is not to say that experience in such locations contributed nothing to the portfolio of images and tropes through which tropicality would later be constructed, as Ptolemy's focus on skin color and megafauna in his discussion of the southern limits of the known world demonstrates. But a glance at a medieval map of the world, such as the Catalan atlas of 1375, suggests that while the scatter of plants and animals across the globe is generally consistent with their actual geographical distributions, they signify a broadly conceived exoticism rather than any thematic mapping of zones. Ethnographic representation makes no distinctions between peoples in terms of dress, political structures, levels of development, or other, later signifiers of tropical difference.

This ontology of the tropics was itself conditioned by similarity and difference within the torrid zone that meshed significantly with European tropical epistemology. Overwhelmingly, the tropical world is maritime. Ptolemy's image of a world dominated by land is truer to the northern temperate zone of the globe than to any other. The further south one moves on the globe, the greater the space occupied by sea rather than land. Access to the tropics required a voyage; thus, for most Europeans, the initial and often principal site of tropical experience was the deck of a ship. The ocean voyage—and its elemental spaces of sea and storm, sun and sky—generates its own experiences of the tropics, whose arrival is more immediately apparent afloat, in a navigational world of astronomical instruments and calculation, than it is on land. Specific on-board skills and rituals punctuated and emphasized the body's own register of a different world: instrumental observation of sun and stars and "crossing the line," for example. And the coast—always much more than the traced line of a map—was predictably the most intense site for shaping tropical experience, often further intensified by such natural phenomena as the dramatic mountain ranges fringing the coasts of Brazil and the coral reefs and lagoons that control access to the islands and atolls of the Caribbean, Pacific, and Indian oceans. Above all, the maritime nature of European access to the tropics placed a premium on vision, in the sense both of visual sweeping of the horizon in anticipation of land and of anticipating the new, the different, the wondrous. These two modes of vision—physical and metaphysical—work constantly to shape each other. From the deck of a ship, tropical land is sensed gradually, the imaginative vision inflected by other senses: the taste of fresh Amazon or Orinoco water diluting the Atlantic or Caribbean brine long before land is seen, the sound of land birds and disappearance of oceanic phenomena, and the olfactory experience of warm offshore breezes.

The more complex vision of a tropical world offered by Humboldt, Bonpland, Hodges, and others and circulated in thematic maps, panoramas, oils, engravings, and lithographs did not diminish the power of maritime experience in shaping tropical views and visions. Continental exploration was still preceded by the sea voyage. The largest part of the tropical world brought within European and American economic space in the nineteenth century was oceanic, through the whale fishery. Writers from Melville to Robert Louis Stevenson ensured that the tropics remained in the popular imagination dominantly as a world of seafaring and island encounters. Afloat, even when mediated through the strongly cognitive practices of sketching, instrumenting, and mapping, the difference of tropical spaces registered itself as directly on the European body as on the minds of Europeans. This remained true into the 1960s when jet travel largely replaced the sea voyage, injecting the visitor directly into tropical space by way of the air-conditioned flight terminal building.

THE ENVIRONMENTAL AND ETHNOGRAPHIC TROPICS

While the mathematical-cosmographic tropics remained a powerful linear trace on the European world map throughout the modern period, sensuous encounter with the radically different physical environments and peoples that lay between Cancer and Capricorn was never completely absent from their mapping. In the *Geography,* Ptolemy himself uses the relationship between the astronomical tropics and an unfamiliar exotic environment as an empirical indicator of the extent of the *oikoumene:* the known, inhabited part of the earth. In the extended critical discussion of the geographer Marinos that occupies much of book 1, Ptolemy points out that his predecessor had placed "the country of the Aithiopians called Agisymba, and Cape Prason, on the parallel that marks the southernmost limit of the known world, and puts this at the Winter Tropic."[3] But Ptolemy is convinced that this is inaccurate. He deploys three sets of arguments for shortening the latitudinal extent of the *oikoumene:* astronomical evidence derived from observation of constellations, the evidence of land journeys and the evidence of sea voyages. The details of his arguments need not detain us, except to note that the case based on the recorded length of land journeys hangs in part on evidence of "where the rhinoceros congregate."[4] The most "manifest" reason to believe Marinos had overestimated the latitudinal extent of the known world combines idealist and empiricist conceptions of the tropics: "The resulting computation brings the Aithiopians and the gathering place of rhinoceros into the frigid zone of the *antoikoumene,* although all animals and plants that are on the same parallel or [parallels] equidistant from either pole ought to exist in similar combinations in accordance with the similarity of their environments."[5] And further:

> This is the [evidence] of the forms and colors of the local animals, from which it would follow that the parallel through the country of Agysimba, which clearly belongs to the Aithiopians, is not as far as the Winter Tropic, but lies nearer the equator. For in the correspondingly situated places on our side [of the equator], that is those on the Summer Tropic, people do not yet have the color of the Aithiopians, and there are no rhinoceros and elephants; but in places not much to the south of these, moderately black people are to be found, such as those who live in the "Thirty Schoinoi" outside Soene. Of the same type too, are the people of the Garame, whom Marinos also says (and indeed for this very reason) live neither right on the Summer Tropic nor to the north, but entirely to the south of it. But in places around Meroe people are already quite black in color, and are at last pure Aithiopians, and the habitat of the elephants and more wonderful animals is there.[6]

The consequence of these observations for Ptolemy is to locate the southern limit of the *oikoumene* at 16°25′ south of the equator, some eight degrees north of the southern tropic. More significant for the present argument, his words demonstrate both a belief in the connection between the cosmographic tropics and a torrid zone characterized by a unique and "wonderful" environment and a belief in a sophisticated geographical discourse that goes beyond a simple coincidence of the two mappings.

The character of the environmental and ethnographic tropics becomes

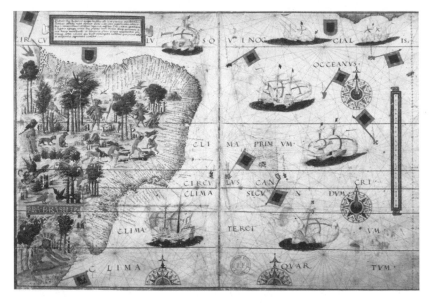

FIGURE 11.1 *Terra Brasilis,* detail from the Miller atlas, 1519 (cliché Bibliothèque nationale de France)

strikingly apparent in mapping and picturing early modern European encounters with the lower latitudes. The European perspective on such spaces was of course almost exclusively maritime; the land expeditions into such latitudes discussed by Ptolemy were not systematically replicated until the late eighteenth century. The sense of marvel and spectacle that informs early written and pictorial accounts of the Caribbean or Calicut owes a debt to medieval European orientalism, to be sure, but it drew, too, on classical motifs, such as the Alexandrine geographer's reference to "elephants and more wonderful animals" and to human skin color as markers of the torrid zone.

The result is captured perfectly in the Miller atlas map of *Terra Brasilis,* drawn in 1519, during the second decade of European contact with tropical South America (fig. 11.1). The parallels marked on the map are not part of a graticule of longitude and latitude used for projection (the map's construction being based on principles of portolan mapping using rhumb lines). They represent instead the classical *klimata* and the southern tropic line. The image is unmistakably "tropical," capturing many of the diverse tropes and elements often attributed to later notions of tropicality. The map is strongly narrative. While the observer is theoretically placed at a vertical distance sufficient to see a continental space stretching longitudinally from the Amazon delta to the northern Pampas, its composition strongly suggests a location for the map reader on board one of the fully rigged Iberian ships sailing toward the coast. And it is the coast exclusively that is named: a tight list of ports, headlands, bays, and other physical features, drawn in the characteristic scalloped curves of the portolan chart. This is a map of coastal progress and possible locations for disembarkation. Penetration beyond the coast is not suggested with any confidence by the map: the mouths of the Amazon and the Plata are strewn with islands, the water inlets between those estuaries end abruptly within the dark pigment used for the coast.

The heavily nominal coast appears in stark contrast to the interior, which, other than the name *Terra Brasilis* is entirely pictorial. The space is composed of scenes of nature, realized in brilliant color. A forested environment of varied and exotic trees, including the signature palm, is alive with exotic fauna: multicolored birds, especially parrots and parakeets, monkeys, and panthers. Human characters are bronzed, mostly naked, and engaged in collecting and stacking wood for shelter or fire, or they present themselves to the observer in elaborate skirts and head dresses of feathers. In both cases, an autochthonous attachment to nature is the dominant ethnographic motif, signifying a fundamental difference from the assumed European observer. In the Miller image, the interior scenes are benign, even Arcadian. Similarly composed renderings of Brazil from subsequent decades and into the seventeenth century present scenes of cannibalism by dwellers within the same environment: the dialectic of "paradise and pestilence" that will henceforth constitute tropicality. The Miller atlas presents to the European vision a place for witness and wonder, not a place to dwell. In short, tropicality is a landscape; a landscape viewed from the imaginative distance offered from on board ship or at best a coastline "civilized" through the tactic of naming or before the framed image of a map or other graphic.

Significantly, the Miller atlas makes no attempt to confine tropical nature within the bounds of the cosmographic tropics; their relationship remains close but casual. Thus a naked Indian kneels among palms on the Pampas of Argentina, nearly twenty degrees south of Capricorn while the map's cartouche describes a single space of monstrous wonder.

TROPICAL MAPPINGS

The imprecise relationship between cosmographic, geographic, and environmental/ethnographic tropics apparent in the Miller atlas map is characteristic of pictorial mapping and descriptive writing throughout the sixteenth and seventeenth centuries, even as the European world map adopted the increasingly "scientific" form signified by accurate plotting of continents and islands against an astronomically correct grid of longitude and latitude lines with tropics, polar circles, and ecliptic clearly delineated. Similar ethnographic scenes to those of the Miller atlas are scattered across the (pseudo) Blaeu *Nova totius terrarum sive novi orbis tabula* of 1665 despite the very clear presence on the map of the great circles. It is, as many of the essays here have stressed, to Alexander von Humboldt that one turns for understanding the effort to map more precisely the connections between the different modes of tropic. In making this claim I am not ignoring the close connections already established, through rival French and British scientific voyages in the Pacific, between astronomical observation, intended in part to determine more precisely the shape of the globe and fixing the longitude at sea, and biogeographical and ethnographic observation, much of it rendered pictorially via the field sketch. Humboldt's work continued and elaborated an already existing project. But he did take it beyond the coastlines that had long confined so much tropical science and experience, and into the continental interior. And, from the perspective of tropical mapping, Humboldt established the significance on the world map of an entirely new form of geographical line, no longer determined astronomically but fixed by empirical observation on the ground and by plotting on the thematic map.

The isoline, famously used by Humboldt to reveal the actual pattern of average annual temperatures across the Western hemisphere, was not exactly his own invention. It had been used earlier in the eighteenth century by Edmund Halley to show lines of equal magnetic deviation. But Humboldt's work had far greater scientific and popular impact in illustrating the value of the isoline as a thematic mapping device in scientific representation, partly because of its remapping of the tropics. In the specific case of his isotherms, Humboldt depicted in precise spatial terms the divergence between actual climatic distributions and the Aristotelian climatic zones. The tropical environment could no longer be assumed to coincide, even theoretically, with a simple torrid zone. The cartographic implications are immediately apparent in Heinrich Berghaus's 1852 *Physical Atlas,* prepared in part to illustrate Humboldt's *Cosmos.* Appearing in the same years that Frederic Church was illustrating Humboldt's tropical world in such intensely colored and dramatic landscape paintings as *The Heart of the Andes* (plate 3), Berghaus presents a severely positivist, scientific world, his finely engraved maps, charts, and graphs depicting myriad aspects of nature: climate, hydrology, animal, plant, and anthropogeography. His world maps adopt a cylindrical projection, centered on the Greenwich meridian and cropped to show a land globe between eighty degrees north and sixty degrees south. The equatorial latitudes are thus off-centered and diminished

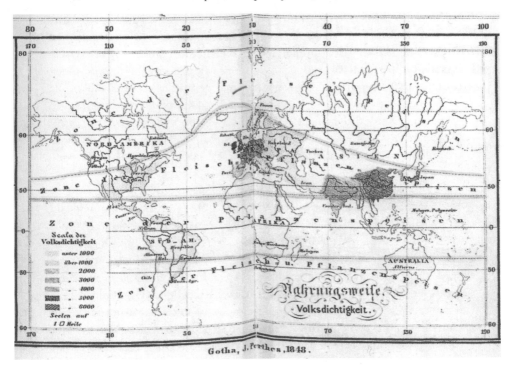

FIGURE 11.2 *Nahrungsweise–Volksdichtigkeit,* insert map in Heinrich Berghaus, *Geographische Verbreitung der Menschen-Rassen* (1852)

relative to the northern temperate and polar latitudes, and while the astronomical great circles are present on the maps, they are indicated indistinctly by very fine lines.

Given that Berghaus shared Humboldt's project of revealing the interrelations of humans within the great scheme of physical nature, his maps naturalize ethnographic materials. He charts the distribution of "human races" principally through text inscriptions across the background world map rather than by means of choropleths. But he illustrates human difference by way of drawn and colored caricatures and skull shapes in the map's margins. The only indication of connection between human types and climatic zones appears in an inset map showing population densities and human diet (fig. 11.2). The world is here divided into belts of flesh eaters, flesh and plant eaters, and plant eaters that correspond to the frigid, temperate, and tropical zones of classical geography. The correlation between plant eaters and the tropics is exact. The shading of human population density in the northern hemisphere suggests graphically that high population densities are to be found only among omnivorous human groups. Cosmographic and environmental tropics, the maps suggest, coincide in precisely the sphere where humans connect most directly with the natural world, in their diet.

Despite masking an unreflective Eurocentrism behind a severely scientific mapping rhetoric, Berghaus's work offers no explicit moral judgments about the various environments and peoples it represents. And its style denotes a clear separation of geographical knowledge from sensuous experience and the romantic constructions of imaginative tropicality becoming popular in the work of artists. This is less apparent in more popular renderings of Humboldtian tropical geography where the convergence of mapping and pictorial imagery seeks to convey a more experiential tropicality within the cool discourse of geographical science. Such an approach is evident in a work directly influenced by Humboldt and Berghaus: *Physical Geography,* by the Swiss-American geographer Arnold Guyot, whose texts were highly influential in American schools and colleges during the late nineteenth century. This book first appeared in 1873 and was revised in 1885, remaining popular in American education into the early years of the twentieth century.[7] Its debt to Humboldt is acknowledged in the opening page engraving of the great man, notebook and pen in hand, cogitating against a background of the Andes. After briefly defining the nature and scope of geography, the first ten pages are devoted to astronomical geography, including a section explaining the "climatic circles." Bizarrely, Guyot harks back to the classical climatic zones, undermining Humboldt's uncoupling of the connection between the parallels and global climatic distributions: "four parallels serve not only to determine position, but also to mark certain important climatic boundaries, hence they may be distinguished as climatic parallels."[8] The section continues by mathematically defining tropics, polar circles, and ecliptic, although nothing further is stated at this point about the connections between these, climate, and landscape. Guyot's later discussion of physical climate corrects the paradox by pointing out the divergence between theoretical climatic bands and thermometrical observation, making direct reference to Humboldt's invention of "isothermals."[9]

When Guyot turns to "life upon earth," however, both text and graphics suggest an intimate connection between latitude and life, with tropicality a recurrent feature, predictably conveyed through striking landscape images. The section's opening engraving depicts "a forest on the Orinoco," its arboreal density luxuriant with palms and lianas, and a distant native canoe negotiating limpid waters strewn with giant water lilies (fig. 11.3). This last part of the work is divided into three substantive sections: "Life in Nature," "Human Life and Social Progress," and "The Human Family." The first of these contains a section devoted precisely to "as-

FIGURE 11.3 *A Forest in the Orinoco,* from Arnold Guyot, *Physical Geography* (New York and Chicago: Ivison, Blakeman, Taylor & Co., 1885)

pects of nature in different zones." It is illustrated by a brightly colored landscape composite, depicting the angle of the summer sun and length of daylight across seventy-five degrees of latitude, marking average mean temperature and iconic cities from Singapore, Madras, Calcutta and Cairo, through New York, Paris, London, and Edinburgh to Saint Petersburg and Hammerfest (plate 9). Sky color, tree and commercial plant forms, and fauna create a panorama graphing the changing natural landscapes of the earth's latitudinal zones. Palm, rhinoceros, elephant, tiger, and monkey signify the tropics, as they had done for Ptolemy. The written description of the tropical zone emphasizes the profusion (even promiscuity), variety, and color of all life forms. Guyot states quite explicitly that "throughout the whole realm of nature, in the animal world as well as in the vegetable, the *development of life* increases in energy, and in the variety and perfection of the types, with increasing intensity of light and heat, from the poles to the tropics."[10]

This claim presents Guyot with a dilemma when he turns finally to "the human family," its characteristics and distributions. The text retails the familiar, if depressing, racial nostrums of its day, with such statements as, "the *form* and *features* show every gradation, from the symmetry, grace, and dignity of the ideal man, portrayed by the sculptors of ancient Greece, to the ugliness and deformity of the Hottentot and the Fuegan."[11] The human races are very clearly ranked in terms of both physical beauty and "perfection" and degree of social and moral "development." But while for every other expression of life on earth Guyot has related such development to latitudinal location, thus privileging the tropics' fecundity, the geographical distribution of humans obeys an alternative spatial logic. "The white race, the normal or typical race," characterized in Guyot's words by "perfect regularity of features, and harmony in all the proportions of the figure, securing agility and strength in the highest degree, with the utmost beauty and grace," occupies a region centered on the mountain lands of Iran and western Asia "where Revelation, the traditions of the nations, and the affinities of their languages alike indicate as the cradle of the human race." [12] Away from this center, human types are gradually modified: "In proportion as we depart from Iran, the geographical center of the races, the regularity of features diminishes, and the harmony of proportions disappears." Guyot notes explicitly this contradiction between his claims about the geographical distribution of life and its relations to the tropics. Humanity is presented as an exception to the general rule: "The law of perfection of type in man, therefore, forms an exception to that observed in the lower orders of creation. The human family appears in its *highest physical perfection,* not within the Tropics, but in the Temperate Zone, in Western Asia, the geographical center of the Old World. . . . The *type degenerates* gradually, with increasing distance, in all directions from this geographical center; until, in the most remote regions of the globe, are found the ugliest, and most deformed specimens of the human family." [13] This claim reworks, in terms of human physiology, a social mapping theorized by conceptual historians and philosophers such as Turgot and Hegel, who tracked human history back in time with increasing geographical distance from Europe. Guyot offers no explanation of the contradiction. Scientific observation of the fecundity and variety of life in tropical regions simply coexists with a continued ethnographic vision of "primitive" peoples, whose state of nature is signified by diet and clothing practices. All is presumably resolved in the book's concluding proclamation "that the entire globe is a grand organism, every feature of which is the outgrowth of a definite plan of an all-wise Creator for the education of the human family, and the manifestation of his own glory." [14]

MAPPING EXPERIENCE: TROPICAL DWELLING

Guyot's tropical geography is by no means unique. Indeed, as a school text, its significance lies precisely in demonstrating how culturally broad and deep was the convergence of Humboldtian science and colonial ideology in framing "tropical space" at the dawn of the twentieth century. Conceptual global synthesis and ethnocentricity generate a sense of tropicality far removed from direct experience of actual places but whose imaginative appeal reached beyond the reveries of American students silent in the February gloom of a Pennsylvania schoolroom. This form of tropicality framed and structured actual lives, in both temperate and tropical latitudes, erasing some of the distinctions among cosmographic, geographic, and environmental tropics. Individuals, however much their lives were shaped by such imaginings, were often required to negotiate a more complex interweaving of ontology and epistemology.

One such life, which continues to shape imaginations and desires for a tropical world even today, was that of the French postimpressionist painter Paul Gauguin. His infant years were spent in tropical Lima and his entire life was an oceanic shuttle between Paris and tropical places: Panama, Martinique, Tahiti. Gauguin was of the same generation as Arnold Guyot, but the painter's geographical imagination and its pictorial expression were shaped by a much more direct and embodied encounter with the different physical, social, and moral environments of the temperate and torrid zones. In contrast to Guyot, for Gauguin, "the increasing intensity of light and heat, from the poles to the tropics" and the fecundity of "tropical nature" found positive expression in the bodies and lives the peoples who inhabited tropical regions.[15] As a painter, he was attracted primarily by the color and atmospheric effects of the higher zenithal angle of sunlight in the tropics. And what Guyot regarded as degeneration from a European peak of civilization, Gauguin and his fellow "primitivists" saw as the furthest remove from the "artificiality" of urban modernity (typified, above all, by Parisian life) and a social innocence, where natural and human life remained in harmonious balance. Having initially sought the natural life in Britanny, his vision moved progressively further from the metropolis. With Vincent van Gogh, Gauguin explored various possible tropical locations in the early 1890s, including Madagascar and Vietnam, for developing a stronger palette and new forms of landscape and for living a primitive life. Their vision was a "tropical workshop" in which European art could be renewed through contact with "a previously unknown, powerfully luminous nature."[16] In the event no "school" materialized; Gauguin lived and worked alone (although never out of contact with the patrons who commissioned his works), first in Tahiti and, as that island came to seem to him too "civilized," in the even more distant Marquesas, where he died in 1903.

Needless to say, the "nature" that Gauguin observed in his tropical dwellings was profoundly humanized, both by indigenous peoples and Europeans. And dreams of harmony and balance between land and life in Polynesia entailed a callous disregard of the decimation wrought by imported diseases on local populations. Gauguin's uncritical participation in

French colonialism (at one point he actively sought employment in the colonial service as a means of financing his tropical existence), his predatory sexuality (in today's terms he was an active pedophile), and his cool financial calculation in choosing locations are undeniable. Similarly, a twenty-first-century eye regards his primitivist philosophy as historically naive and hopelessly compromised. What remains, however, is his belief that light, color, and forms of life do take on specific characteristics in such places, in marked contrast to Western Europe. Gauguin lived nearly half of his life in tropical regions of the world, much of the time as distant from metropolitan influences as any anthropologist, and in intimate physical contact with the natural and social patterns of daily life. The "tropics," in specific locales—Peru, Panama, Polynesia—were for him an experiential reality, a dwelling place, whose effects were registered at a much deeper level than representation alone. A tropical geography of ontological force complicates the theoretical trajectory of Gauguin's art, above all his finished paintings. Choices of subject matter, composition, and palette in Gauguin's Polynesian canvases all speak of a complex meeting of European aesthetics with land and life in the South Pacific. As Jean-François Staszak has shown, Gauguin's actions and his letters betray a growing recognition, too, of the fraught connections and tensions between the primitivist dream and resistant materialities of land and life in French Polynesia. Like European painters before and after him, Gauguin pursued effects of light, atmosphere and color, and a physical intensity of organic life that constitute environmental actualities, and his canvases are among the most successful in mapping such spaces.

If Gauguin captures visually the intercourse of temperate longings and tropical dwelling, another Frenchman, who experienced the twilight of his country's colonial experience in the tropics, captures it textually. Claude Lévi-Strauss's 1955 account of study in the Brazilian interior is much more than an ethnographic narrative or even an empirical foundation for his enormously influential structuralist version of anthropology. Its title, *Tristes tropiques,* is hauntingly evocative, echoing a sweet nostalgia for prewar Paris and for the writer's student days, as much as for the bodily sensations of an unfamiliar world—the warmth, smells, colors, and tastes of Brazil—and for a haunting primitivism not wholly dissimilar to Gauguin's. While field reports of his time among various Brazilian Indian groups form the core of his book, more than passing reference is made to the author's experiences in India, Egypt, Indo-China, and other tropical regions of the globe. But he avoids any grand geographic synthesis, and his text is structured as a spatial narrative, originating in Paris and returning, in the final chapter, to the metropolis as a place for reflection and critique.

Read critically it is clear that *Tristes tropiques* mobilizes virtually every literary and geographic trope associated with tropicality: the ocean voyage, the dawning sense of difference, the foliant luxury of the coastal view, the sensuality of first encounter, the emotional heft of dawn and dusk in the lower latitudes, and always primitiveness, paradise, and pestilence. Like Gauguin's, this is a journey to the edges of space and the beginnings of time, precisely in the traditions of Turgot and Hegel. Pages are devoted to the *longeurs* of the ocean voyage and the psychological impacts of space and time, elements and heavens. By the time he enters the tropics, the author seems already to be proclaiming himself a changed being, and the length of his time in Brazil confirms him—to himself at least—as dweller. A deep moral gulf is signified by changes in the physical world. But Lévi-Strauss's use of the conventions of tropicality is knowing; they permit a critical reflection on the epistemological assumptions that shape his narrative, composed long after the event and through the lens of his long stay in tropical Brazil:

> The inky sky over the Doldrums and the oppressive atmosphere are more than just an obvious sign of the nearness of the equator. They epitomize the moral climate in which two worlds have come face to face. The cheerless sea between them, and the calmness of the weather whose only purpose seems to be to allow evil forces to gather fresh strength, are the last mythical barrier between two regions so diametrically opposed to each other through their different conditions that the first people to become aware of the fact could not believe that they were both equally human. A continent barely touched by man lay exposed to men whose greed could no longer be satisfied by their own continent. Everything would be called into question by this second sin: God, morality and law . . . everything would be verified in practice and revoked in principle: the Garden of Eden, the Golden Age of antiquity, the Fountain of Youth, Atlantis, the Hesperides, the Islands of the Blessed, would be found to be true; but revelation, salvation, customs and law would be challenged by the spectacle of a purer, happier race of men (who, of course, were not really purer or happier, although a deep-seated remorse made them appear so). Never had humanity experienced such a harrowing test.[17]

The material presence of the tropical earth impresses itself bodily, and in ways that David Arnold has described as characteristic of tropical narratives. Yet Lévi-Strauss insists that representations of the tropics had left him quite unprepared for his physiological response:

> The traveler approaching the New World is first conscious of it as a scent very different from the one suggested back in Paris by the connotations of the word Brazil, and very difficult to describe to anyone who has not

experienced it. At first, it seemed that the sea smells of the preceding weeks had ceased to circulate so freely; they had come up against an invisible wall: thus immobilized, they no longer claimed the traveler's attention, which was now drawn towards smells that were of a quite different nature and that nothing in his past experience enabled him to define: they were like a forest breeze alternating with hot-house scents, the quintessence of the vegetable kingdom, and held a peculiar freshness so concentrated as to be transmuted into a kind of olfactory intoxication.[18]

And a little later, after leaving Rio de Janeiro for Santos: "As the boat slowly made its way between the islands, for the first time I experienced the impact of the tropics. We were encompassed by a channel of greenery. It was as if we only had to stretch out our hands to grasp the plants."[19]

The ethnographic accounts that constitute the major part of the text certainly trace a predictable path, from the immediate environs of São Paulo, where Lévi-Strauss initially believed he could understand Brazil's pre-Columbian peoples, through Mato Grosso and into the Amazon basin. Descriptions of nature, landscape, and life—now living and traveling with native peoples—oscillate between poles: barren and fertile, "fantastic garden" and "bleak savannah," adventure and boredom.[20] But throughout the text, there is an insistent return to the direct impress of this world on human bodies, both the author's and those of the Indians he is accompanying. Of course, this, too, is open to interpretation as further evidence of "naturalizing" the tropics as "other" to the equable nature and cerebral existence of the normalized temperate zone. But, as in Gauguin's case and that of all those who actually engaged rather than merely reported the tropics, this does insufficient justice to the ontological challenge of Lévi-Strauss's physical encounter. Only through the careful observation and honest interrogation across difference can we possibly escape the solipsism of sameness. Lévi-Strauss's own struggle with the intersection of experience and representation in retailing tropical experience is addressed directly toward the end of the book:

But if the inhabitants were mute, perhaps the earth itself would speak to me. Over and above the marvels which had enchanted me along the river, perhaps it would answer my prayer and let me into the secret of its virginity. Where exactly does that virginity lie, behind the confusion of appearances which are all and yet nothing? I can pick out certain scenes and separate them from the rest; is it this tree, this flower? They might well be elsewhere. Is it also a delusion that the whole should fill me with rapture, while each part, taken separately, escapes me? If I have to accept it as being real, I want at least to grasp it in its entirety, down to its last constituent element. I reject this vast landscape, I circumscribe it and reduce it to this clayey beach and this blade of grass; there is nothing to prove that my eye, if it broadened its view of the scene, would not recognize the Bois de Meudon around this insignificant fragment, which is trodden by the most authentic savages but from which, however, Man Friday's footprint is missing.[21]

CONCLUSION

My distinction of three "ontological tropics"—cosmographic, geographic, and environmental/ethnographic—is principally heuristic. It is intended as a device to explore the complex interplay of epistemologies with spaces generated by empirical and idealist framings of the tropics. I have suggested that tropicality is a landscape, one that draws—to be sure—on these different expressions of tropical space. As a landscape, tropicality is at once phenomenological and representational, working through various acts of mapping. Examples from Ptolemy, through the makers of the Miller atlas to Guyot, Gauguin, and Lévi-Strauss complicate any simple relations between the ontological tropics and tropicality as an epistemology. Physical encounter draws on the representational tropes of mapping, certainly, but moves beyond these to confront tropical places as habitus. More recently, the artist Marc Dion has foregrounded this complexity of "tropical." In his 1991 site-specific project, *On Tropical Nature,* he selected a paradigmatic tropical site: an uninhabited spot in the Orinoco rain forest, where he camped for three weeks, collecting plant, animal, and other specimens. These were crated and delivered each week to a room in a Caracas art institution where they were later unpacked and displayed as art works and contextualized within the curatorial framework of this second site. This conceptual work insists on the materiality of tropical nature while making it representational within the context of the classifying, mapping, and viewing practices that have conventionally framed the tropics.[22]

Somewhere between the theoretical mapping that inscribes cosmographic lines around the imaged globe to produce "the tropics," the empirical geography of lands, seas, and airs that characterize these parts of the globe, and the environments and peoples that make places in the equinoctial regions of the globe, epistemology and ontology constantly rework one another. While not escaping "visions" entirely, it seems at least possible to transcend purely cognitive geographies toward a more sensual if not visceral engagement with actual places. Where, when, and how the material world and the ways we come to know it come together is not uniform across specific instances. The essays in this volume variously capture and interpret many of those instances, moments of European and

American "envisioning" and representation of the tropical world in the period of European tropical colonialism. Their arrangement into voyages, mappings, and sites traces narratively not only the characteristic ways that Europeans came into physical contact with the new worlds lying between Cancer and Capricorn but also the always anxious, unnerving, and muddied intersection between what is there in the world—immediately present to the sensing and social body—and how it can be known consciously and represented honestly. Like any other geography, the tropics is a place, made up of places that are tropical in the myriad ways that humans inhabit them.

In the opening paragraphs of this chapter, I alluded to O. H. K. Spate's words about the Pacific and his distinction between its existence as a European artifact and as made by Pacific peoples who are themselves artificers. Reading *Tropical Visions* and recognizing the fuzziness of boundaries between geographic ontologies and the various mapping epistemologies that I have tried to explore here, we might wish to modify this neat distinction. "The tropics" are a cosmographic effect, connecting earth and heavens: a European artifact, to be sure, but, as Guyot's simple geography lesson makes clear, the movements of sun and earth and the nature of organic existence mean that the physical world is also an artificer. As Humboldt himself might have remarked, the diversity in experiences, communication, and representations of the geographies that exist in the spaces between Cancer and Capricorn comes into sharper focus when we do not neglect the unities through which they are woven.

Notes

CHAPTER I

1. Felix Driver, "Imagining the Tropics: Views and Visions of the Tropical World," *Singapore Journal of Tropical Geography* 25, no. 1 (2004): 1–17.

2. Omar Ribeiro Thomaz, "Do Saber Colonial ao Luso-Tropicalismo: 'Raça' e 'Nação' nas Primeiras Décadas do Salazarismo," in *Raça, Ciência e Sociedade,* ed. Marcos Chor Maio and Ricardo Ventura Santos (Rio de Janeiro: Fiocruz/CBBB, 1996); David Arnold, "Inventing Tropicality," in *The Problem of Nature: Environment, Culture and European Expansion* (Oxford: Blackwell, 1996), 161–62.

3. Nancy Leys Stepan, "Tropical Modernism: Designing the Tropical Landscape," *Singapore Journal of Tropical Geography* 21, no. 1 (2000): 76–91.

4. C. Dunn, "Tropicália, Counterculture and the Diasporic Imagination in Brazil," in *Brazilian Popular Music and Globalization,* ed. C. A. Perrone and C. Dunn (New York: Routledge, 2002).

5. Compare F. R. Aparicio and S. Chávez-Silverman, eds., *Tropicalizations: Transcultural Representations of Latinidad* (Hanover, NH: University Press of New England, 1997); and Mary-Louise Pratt, *Imperial Eyes: Travel Writing and Transculturation* (London: Routledge, 1992).

6. Arnold, "Inventing Tropicality"; Gavin Bowd and Daniel Clayton, "Fieldwork and Tropicality in French Indochina: Reflections on Pierre Gourou's *Les paysans du delta tonkinois,* 1936," *Singapore Journal of Tropical Geography* 24, no. 2 (2003): 147–68.

7. See, among others, Nicholas Thomas, *Colonialism's Culture: Anthropology, Travel and Government* (Oxford: Polity, 1994).

8. A fascinating, if relatively minor, example of this process at work may be found in the career of historian Fernand Braudel, who like several of his peers (including the anthropologist Claude Lévi-Strauss and the geographer Pierre Monbeig) spent a formative period teaching in Brazil in the 1930s. Braudel once remarked that it was his period in Brazil that turned him into a true intellectual. It appears that the design of that master work of European history, *The Mediterranean and the Mediterranean World in the*

Age of Philip II, was conceived not in France but in the very nearly tropical São Paulo—or rather, in the transatlantic shuttling between these worlds that Braudel experienced during the late 1930s, wintering in the archives of Europe, spending the rest of the year in Brazil. See E. Paris,"L'époque brésilienne de Fernand Braudel (1935– 1937)," *Storia della Storiografia* 30 (1996): 56–68; T. E. Skidmore, "Lévi-Strauss, Braudel and Brazil: A Case of Mutual Influence," *Bulletin of Latin American Research* 22 (2003): 340–49.

9. See, for example, Deborah Poole, *Vision, Race and Modernity: A Visual Economy of the Andean Image World* (Princeton, NJ: Princeton University Press, 1997); Christopher Pinney and Nicolas Peterson, eds., *Photography's Other Histories* (Durham, NC, and London: Duke University Press, 2003).

10. See Srinivas Aravamudan, *Tropicopolitans: Colonialism and Agency, 1688–1804* (Durham, NC, and London: Duke University Press, 1999).

11. Humboldt, cited in Michael Dettelbach's chapter in this volume. See also Dorinda Outram, "On Being Perseus: New Knowledge, Dislocation and Enlightenment Exploration," in *Geography and Enlightenment,* ed. David N. Livingstone and Charles W. J. Withers (Chicago: University of Chicago Press, 1999).

12. Darwin, cited in S. F. Cannon, *Science in Culture: The Early Victorian Period* (New York: Science History Publications, 1978), 87. See also Luciana Martins, "A Naturalist's Vision of the Tropics: Charles Darwin and the Brazilian Landscape," *Singapore Journal of Tropical Geography* 21, no. 1 (2000): 19–33.

13. Felix Driver and Luciana Martins, "John Septimus Roe and the Art of Navigation," *History Workshop Journal* 54 (2002): 144–61; Paul Carter, "Dark with Excess of Bright: Mapping the Coastlines of Knowledge," in *Mappings,* ed. Denis Cosgrove (London: Reaktion Books, 1999), 125–47; D. G. Burnett, *Masters of All They Surveyed: Exploration, Geography and a British El Dorado* (Chicago: University of Chicago Press, 2000), esp. 67–117.

14. Michael Dettelbach, "Global Physics and Aesthetic Empire: Humboldt's Physical Portrait of the Tropics," in *Visions of Empire: Voyages, Botany and Representations of Nature,* ed. D. P. Miller and P. H. Reill (Cambridge: Cambridge University Press, 1996); M. Nicholson, "Alexander von Humboldt and the Geography of Vegetation," in *Romanticism and the Sciences,* ed. A. Cunningham and N. Jardine (Cambridge: Cambridge University Press, 1990); Nigel Leask, "Alexander von Humboldt and the Romantic Imagination of America: The Impossibility of Personal Narrative," in *Curiosity and the Aesthetics of Travel Writing, 1770–1840* (Oxford: Oxford University Press, 2002), esp. 246–56.

15. Leask, *Curiosity and the Aesthetics of Travel Writing,* 248–49.

16. This argument is briefly made in Felix Driver and Brenda Yeoh, "Constructing the Tropics," *Singapore Journal of Tropical Geography* 21 (2000): 1–5. More generally, see Nicholas Thomas, *Entangled Objects: Exchange, Material Culture and Colonialism in the Pacific* (Cambridge, MA: Harvard University Press, 1991).

17. Johannes Fabian, *Out of Our Minds: Reason and Madness in the Exploration of Central Africa* (Berkeley: University of California Press, 2000).

18. Denis Cosgrove, *Apollo's Eye: A Cartographic Genealogy of the Earth in the Western Imagination* (Baltimore: Johns Hopkins University Press, 2001), 29–53.

19. Ibid., 130–38.

20. Stephen Greenblatt, *Marvellous Possessions: The Wonder of the New World* (Oxford: Oxford University Press, 1991); Hugh Raffles, *In Amazonia: A Natural History* (Princeton, NJ, and Oxford: Princeton University Press, 2002), esp. 95–110; Candace Slater, "El Dorado and the Golden Legacy," in *Entangled Edens: Visions of the Amazon* (Berkeley: University of California Press, 2002).

21. The term "planetary consciousness" is used by Mary-Louise Pratt in *Imperial Eyes*. See also Matthew Edney, *Mapping an Empire: The Geographical Construction of British India, 1765–1843* (Chicago: University of Chicago Press, 1997); Miller and Reill, eds., *Visions of Empire;* Robert Stafford, *Scientist of Empire: Sir Roderick Murchison, Scientific Exploration and Victorian Imperialism* (Cambridge: Cambridge University Press, 1989); Richard Drayton, *Nature's Government: Science, Imperial Britain, and the "Improvement" of the World* (New Haven, CT: Yale University Press, 2000); Charles W. J. Withers, "Geography, Natural History and the Eighteenth-Century Enlightenment: Putting the World in Place," *History Workshop Journal.* 39 (1995): 137–63.

22. Bernard Smith, *European Vision and the South Pacific,* 2d ed. (New Haven, CT: Yale University, 1985). The book was first published in 1960.

23. The art of William Hodges, central to Smith's argument, is the subject of a notable exhibition recently curated by Geoff Quilley. Smith's work provides a fundamental reference point in the accompanying book, though its conceptual framework is questioned: see Geoff Quilley and John Bonehill, eds., *William Hodges, 1744–1797: The Art of Exploration* (New Haven, CT: Yale University Press, 2004).

24. For reworkings of the theme of encounter and visual culture, see the excellent collection by Nicholas Thomas and Diane Losche, eds., *Double Vision: Art Histories and Colonial Histories in the Pacific* (Cambridge: Cambridge University Press, 1999).

25. Felix Driver, "Distance and Disturbance: Travel, Exploration and Knowledge in the Nineteenth Century," *Transactions of the Royal Historical Society,* 6th ser., 14 (2004): 73–92.

26. Jonathan Lamb, *Preserving the Self in the South Seas, 1680–1840* (Chicago: University of Chicago Press, 2001), 7.

27. Dorinda Outram, "On Being Perseus" and "New Spaces in Natural History," in *Cultures of Natural History,* ed. Nicholas Jardine, James Secord, and Emma Spary (Cambridge: Cambridge University Press, 1996), 249–65.

28. Livingstone and Withers, eds., *Geography and Enlightenment.*

29. See, esp., Denis Cosgrove, "Introduction: Mapping Meaning," in *Mappings,* ed. Cosgrove, 1–23.

30. S. F. Cannon, "Humboldtian Science," in *Science in Culture: The Early Victorian Period* (New York: Science History Publications, 1978), 92.

31. Nicholson, "Alexander von Humboldt," 173–78; Dettelbach, "Global Physics and Aesthetic Empire," 267–72.

32. Michael Dettelbach, "The Face of Nature: Precise Measurement, Mapping and Sensibility in the Work of Alexander von Humboldt," *Studies in the History and Philosophy of the Biological and Biomedical Sciences* 30 (1999): 487, 490.

33. Michael Bravo, "Ethnological Encounters," in *Cultures of Natural History,* ed. Jardine, Secord and Spary, 347.

34. Michael Bravo, "Precision and Curiosity in Scientific Travel: James Rennell and the Orientalist Geography of the New Imperial Age (1760–1830)," in *Voyages and*

Visions: Towards a Cultural History of Travel, ed. J. Elsner and J.-P. Rubiés (London: Reaktion, 1999), 162–83.

35. On the role of comparative methods of map compilation, see Edney, *Mapping an Empire;* on local reliance on Indian surveying skills, see Kapil Raj, "Circulation and the Emergence of Modern Mapping: Great Britain and Early Colonial India, 1764–1820," in *Society and Circulation: Mobile People and Itinerant Cultures in South Asia, 1750–1950,* ed. C. Markovits, J. Pouchepadass, and S. Subrahmanyam (Delhi: Permanent Black, 2003), 23–54.

36. Nicholas Thomas, "'On the Varieties of the Human Species': Forster's Comparative Ethnology," in J. R. Forster, *Observations Made during a Voyage Round the World,* ed. N. Thomas, H. Guest, and M. Dettelbach (Honolulu: University of Hawaii Press, 1996), xxxii.

37. See, for example, Bronwen Douglas, "Art as Ethno-Historical Text: Science, Representation and Indigenous Presence in Eighteenth and Nineteenth-Century Oceanic Voyage Literature," in *Double Vision,* ed. Thomas and Losche.

38. Peter Hulme, *Colonial Encounters: Europe and the Native Caribbean, 1492–1797* (London: Routledge, 1986).

39. David Mackay, *In the Wake of Cook: Exploration, Science and Empire, 1780–1801* (London: Croom Helm, 1985); Richard Grove, *Green Imperialism: Colonial Expansion, Tropical Island Edens and the Origins of Environmentalism, 1600–1860* (Cambridge: Cambridge University Press, 1995), 339–40; Miller and Reill, eds., *Voyages of Empire,* esp. chapters by Mackay and Frost.

40. On visual culture and the circulation of ideas and images of race across the Atlantic world, see Geoff Quilley and Kay Dian Kriz, eds., *An Economy of Colour: Visual Culture and the Atlantic World, 1660–1830* (Manchester: Manchester University Press, 2003).

41. Arnold, "Inventing Tropicality"; Nancy Leys Stepan, *Picturing Tropical Nature* (London: Reaktion, 2001); Katherine Emma Manthorne, *Tropical Renaissance: North American Artists Exploring Latin America, 1839–1879* (Washington, DC: Smithsonian Institution Press, 1989); P. Diener and M. F. Costa, *Rugendas e o Brasil* (São Paulo: Capivara, 2002).

42. Jane Camerini, "Evolution, Biogeography, and Maps: An Early History of Wallace's Line," *Isis* 84 (1993): 700–727; Nicholas Thomas, "Melanesians and Polynesians: Ethnic Typifications Inside and Outside Anthropology," in *In Oceania: Visions, Artifacts, Histories* (Durham, NC: Duke University Press, 1997), 133–55.

43. David Arnold, "India's Place in the Tropical World, 1770–1930," *Journal of Imperial and Commonwealth History* 26 (1998): 6–9. On subsequent interpretations of "jungle," see T. G. Birtles, "First Contact: Colonial European Preconceptions of Tropical Queensland Rainforest and Its People," *Journal of Historical Geography* 23 (1997): 393–417; M. Sioh, "Authorizing the Malaysian Rainforest: Configuring Space, Contesting Claims and Conquering Imaginaries," *Ecumene* 5 (1998): 144–66; Slater, *Entangled Edens.*

44. Luciana Martins, "Mapping Tropical Waters," in *Mappings,* ed. Cosgrove, 148–68.

45. Luciana Martins, "The Art of Tropical Travel, 1768–1830," in *Georgian Geographies,* ed. Miles Ogborn and Charles W. J. Withers (Manchester: Manchester

University Press, 2004), 72−91; Cecilia Powell, *Italy in the Age of Turner: "The Garden of the World"* (London: Dulwich Picture Gallery, 1998).

46. David N. Livingstone, "Tropical Climate and Moral Hygiene: The Anatomy of a Victorian Debate," *British Journal for the History of Science.* 32 (1999): 93−100; Mark Harrison, "Tropical Medicine in Nineteenth-Century British India," *British Journal for the History of Science* 25 (1992): 299−318; and R. MacLeod and M. Lewis, eds., *Disease, Medicine and Empire: Perspectives on Western Medicine and the Experience of European Expansion* (London: Routledge, 1988). The discourse on acclimatisation was already well-established as a "paradigmatic colonial science"; see, esp., Michael Osborne, "Acclimatizing the World," in "Nature and Empire: Science and the Colonial Enterprise," ed. MacLeod, Roy, *Osiris* 15 (2001): 135−51.

47. S. Eisenman, *Gauguin's Skirt* (London: Thames & Hudson, 1997), 54.

48. Travel brochure, *Exotic Holidays* (London: Sunset, n.d.), 133.

CHAPTER 2

This chapter is a substantially revised version of "On the spot: L'artista-viaggiatore e l'inventario iconografico del mondo (1772−1859)," *Geotema* 3, no. 8 (1997): 137−49. The translation is by the volume's editors. The dates in the title refer to Captain Cook's first Pacific voyage, which coincides with the birth of Alexander von Humboldt in 1769, followed by the date of Humboldt's death, which coincides with the publication of Charles Darwin's *On the Origin of Species,* in 1859. The epigraph is from Richard Keynes, ed., *The Beagle Record* (Cambridge: Cambridge University Press, 1979), 169.

1. The term, as used here, has no ready equivalent in Italian (*dal vero? sul campo?*).

2. See Bernard Smith, *European Vision and the South Pacific,* 2d ed. (New Haven, CT, and London: Yale University Press, 1985). See also Barbara Maria Stafford, *Voyage into Substance: Art, Science, Nature and Illustrated Travel Account, 1760−1840* (Cambridge, MA, and London: MIT Press, 1984); Michael Jacobs, *The Painted Voyage: Art, Travel and Exploration, 1564−1875* (London: British Museum, 1995); and F. Moureau, ed., *L'oeil aux aguets ou l'artiste en voyage* (Paris: Klincksiek, 1995).

3. William Hodges, *Select Views in India. Drawn on the Spot in the Years 1780−1783, and Executed in Aquatint* (London: Edwards, 1786).

4. James Cook, *A Voyage toward the South Pole, and Round the World . . . Illustrated with Maps, Charts, a Variety of Portraits of Persons, Views of Places Drawn during the Voyage by Mr. Hodges, and Engraved by the Most Eminent Masters,* 2 vols. plus atlas (London: Strahan & Cadell, 1777).

5. Alexander von Humboldt, *Cosmos: A Sketch of the Physical Description of the Universe,* 2 vols (Baltimore and London: Johns Hopkins University Press, 1997), 2:97.

6. Georg Forster, *A Voyage Round the World* (London: White, Robson, Elmsly, and Robinson, 1777), 427−28, quoted in Smith, *European Vision and the South Pacific,* 74−75.

7. Rüdiger Joppien and Bernard Smith, *The Art of Captain Cook's Voyages,* 4 vols. (Melbourne: Oxford University Press, 1985).

8. John Hawkesworth, *An Account of the Voyages Undertaken by the Order of his Present Majesty for Making Discoveries in the Southern Hemisphere,* 3 vols. (London: Strahan & Cadell, 1773).

9. S. Parkinson, *A Journal of a Voyage in the South Seas...*, *Faithfully Transcribed from the Papers of the Late S. P., Draughtsman to Joseph Banks . . . Embellished with Views, and Designs, Delineated by the Author, and Engraved by Capital Artists* (London: Stanfield Parkinson, 1773).

10. The phrase "Everything that it is impossible to describe" is used by La Pérouse on the first page of his journal, with reference to Duché de Vancy.

11. Smith, *European Vision and the South Pacific*, 109.

12. J. Webber, *Views of the South Seas* (London: Boydell, 1808).

13. George Vancouver, *A Voyage of Discovery to the North Pacific Ocean and round the World . . . (1790−95)*, 3 vols. (London: Robinson & Edwards, 1798).

14. J. Cook and J. King, *A Voyage to the Pacific Ocean*, 3 vols. plus atlas (London: Strahan, 1784); J. F. de Galaup de La Pérouse, *Voyage autour du monde . . . rédigé par M. L. A. Milet-Mureau.* 4 vols. plus atlas (Paris: Imprimerie de la République, 1797).

15. Jacques Gérard Milbert (1766−1850) was a pupil of Pierre-Henry de Valenciennes, one of the greatest landscape painters in the age of Grand Tour. Milbert would eventually publish a series of views: J. G. Milbert, *Voyage pittoresque à l'Ile de France, au Cap de Bonne Espérance et à l'Ile de Ténériffe,* 2 vols. plus atlas (Paris: Nepveu, 1812).

16. Mildred Archer, *Early Views of India: The Picturesque Journey of Thomas and William Daniell, 1786−1794* (London: Thames & Hudson, 1980).

17. R. Home, *Select Views in Mysore* (London: Bowyer, 1794).

18. T. Daniell, *Hindoo Excavations in the Mountain of Ellora . . . in Twenty-four Views from the Drawings of James Wales* (London: Longman, Hurst, Rees, Orme & Brown, 1816). See also J. Wales, *Hindoo Excavations in the Mountain of Ellora near Aurungabad in the Deccan: In Twenty-four Views,* 2 vols. (London: Thomas Daniell, 1803−4), and *Bombay Views: Twelve Views of the Island of Bombay and Its Vicinity, Taken in the Years 1791 and 1792* (London: Goodwin, 1800).

19. For example, T. Daniell and W. Daniell, *Antiquities of India: Twelve Views from the Drawings* (London: printed by the authors, 1799); W. Daniell, *Interesting Selection from Animated Nature with Illustrative Scenery,* 3 vols. (London: Cadell & Davies, 1808); T. Daniell and W. Daniell, *A Picturesque Voyage to India by Way of China* (London: Longman, Hurst, Rees, Orme, 1810); W. Daniell, *Illustrations of the Island of Staffa, in a Series of Views* (London: Longman, Hurst, Rees, Orme, 1818).

20. T. Daniell and W. Daniell, *Oriental Scenery,* 6 vols. (London: Longman, Hurst, Rees, Orme, 1795−1815).

21. Peter Bowler, *The Earth Encompassed: A History of the Environmental Sciences* (New York and London: W. W. Norton & Co., 2000).

22. Martin Kemp, *The Science of Art: Optical Themes in Western Art from Brunelleschi to Seurat* (New Haven, CT, and London: Yale University Press, 1990), 221.

23. Smith, *European Vision and the South Pacific*, 65.

24. J. P. Hackert, "Due lettere sulla pittura di paesaggio," in *Il paesaggio secondo natura: Jacob Philip Hackert e la sua cerchia,* ed. P. Chiarini (Rome: Artemide, 1994), 310−27.

25. W. Hamilton, *Campi Phlegraei: Observations on the Volcanos of the Two Sicilies,* 2 vols. (Naples, 1776).

26. R. Hentzi, *Vues remarquables des montagnes de la Suisse* (Bern: Wagner, 1776).

27. C. Wolf, *Vues remarquables des montagnes de la Suisse avec leur description* (Bern:

Wagner, 1778), and *Vues remarquables des montagnes de la Suisse* (Paris, 1787−91); C. M. Decourtis, *Vues remarquables des montagnes de Suisse* (Amsterdam: Yntema, 1785).

28. M. T. Bourrit, *Nouvelle description des vallés de glace et des hautes montagnes qui forment la chaîne des Alpes Pennines et Rhétiennes,* 3 vols. (Geneva: P. Barde, 1783).

29. H. B. Saussure, *Voyage dans les Alpes,* 4 vols. (Neuchatel, 1779).

30. J. W. Goethe, *Italian Journey [1786−1788],* trans. W. H. Auden and E. Mayer (London: Penguin Books, 1962), 278, 304.

31. Ibid., 233. See J. C. R. de Saint-Non, *Voyage pittoresque ou description des royaumes de Naples et Sicile,* 5 vols. (Paris: Clousier, 1781−86).

32. Goethe, *Italian Journey,* 220.

33. J. Houel, *Voyage pittoresque des isles de Sicile, de Malte et de Lipari,* 4 vols. (Paris: Imprimerie de Monsieur, 1782).

34. J. B. de Laborde, *Voyage pittoresque de la France,* 8 vols. (Paris: Imprimerie de Monsieur, 1781−96).

35. J. B. de Laborde, *Tableaux topographiques, pittoresques, physiques, historiques, moraux, politiques, littéraires de la Suisse et d'Italie, ornés de 1.200 estampes* (Paris: Née & Masquelier, 1786).

36. M. G. A. F. de Choiseul-Gouffier, *Voyage pittoresque de la Grèce,* 2 vols. (Paris: Tilliard, De Bure, 1782, 1809); L. F. Cassas, *Voyage pittoresque de la Syrie, de la Phénicie, de la Palestine, et de la Basse Egypte* (Paris: Imprimerie de la République, 1798−99), and *Voyage pittoresque et historique de l'Istrie et de la Dalmatia* (Paris: Didot, 1801); L. J. de Laborde, *Voyage pittoresque et historique de l'Espagne.* 4 vols. (Paris: Didot, 1806−18); J. G. Milbert, *Voyage pittoresque à l'Ile de France, au Cap de Bonne Espérence et à l'Ile de Ténériffe,* 2 vols. plus atlas (Paris: Nepveu, 1812).

37. F. Skjöldebrand, *Voyage pittoresque au Cap Nord* (Stockholm: C. Deleen & J. D. Forsgren, 1801−2); F. Fontani, *Viaggio pittorico della Toscana* (Florence: Tofani, 1801−3).

38. J. B. Debret, *Voyage pittoresque et historique au Brésil depuis 1816 jusqu'à 1831,* 3 vols. (Paris: Firmin-Didot, 1834).

39. A. von Humboldt, *Vues des Cordillères et monumens des peuples indigènes de l'Amérique* (Paris: F. Schoell, 1810).

40. Humboldt, *Cosmos,* 2:93−94.

41. R. Brown, *Prodomus Florae Novae Hollandiae* (London: R. Taylor & Socii, 1810).

42. M. Flinders, *A Voyage to Terra Australis, . . . in the Years 1801−03, in His Majesty's Ship the Investigator,* 2 vols. plus atlas (London: Bulmer, 1814).

43. Smith, *European Vision and the South Pacific,* 195.

44. W. E. Parry, *Journal of a Voyage for the Discovery of the North-West Passage* (London: Murray, 1821). See also W. Westall, *Views of the Caves Near Ingleton . . . in Yorkshire* (London: Murray, 1818).

45. O. von Kotzebue, *Entdeckungs-Reise in die Süd-See und nach der Berings-Strasse* (Weimar: Hoffmann, 1815−18).

46. L. Choris, *Voyage pittoresque autour du monde* (Paris: Firmin-Didot, 1820).

47. L. Choris, *Vues et paysages des régions équinoxiales* (Paris: Renouard, 1826).

48. For a reproduction of this image, see Douglas Botting, *Humboldt and the Cosmos* (London: Joseph, 1973), 271.

49. Carl Gustav Carus, *Neun Briefe über Landschaftsmalerei* (Leipzig, 1831).

50. Keynes, ed., *The Beagle Record,* 169. See also Robert Fitzroy, *Narrative of the Surveying Voyages of His Majesty's Ships Adventure and Beagle, between the Years 1826 and 1836,* 3 vols. plus atlas (London: Colburn, 1839).

51. Augustus Earle, *Sketches Illustrative of the Native Inhabitants and Islands of New Zealand* (London. R. Martin & Co., 1838).

52. Quoted by Smith, *European Vision and the South Pacific,* 252.

53. Humboldt, *Cosmos,* 2:97.

CHAPTER 3

1. Alexander von Humboldt, *Reise auf dem Rio Magdalena, durch den Anden und Mexico* (Berlin: Beiträge zur Alexander-von-Humboldt-Forschung, 1986), 128–35, excerpted in Alexander von Humboldt, *Vues des Cordillères et monumens des peuples indigènes de l'Amérique* (Paris, 1810), 16–17. Note that this is not what Humboldt has chosen to depict in his drawing of the traverse. Here he contrasts the muscular figures of the porters with the overrefined Creole, whose feet never touch the ground and who reads a book. The contrast between aristocratic overrefinement, produced by caste isolation and lack of sociability or fellow feeling, and the true humanity of the explorer is emphasized here.

2. Humboldt to Joseph Banks, 20 June 1797, and Humboldt to David Friedländer, 11 April 1799, in Fritz Lange and Ilse Jahn, eds., *Die Jugendbriefe Alexander von Humboldts* (Berlin: Akademie Verlag, 1973), 584, 657.

3. Alexander von Humboldt and Aimé Bonpland, *Plantes équinoxiales,* 2 vols. (Paris: F. Schoell, 1808–9, reprint, Amsterdam: Theatrum Orbis Terrarum, 1971), 1:17–18.

4. Humboldt to Karl Ludwig Willdenow, Havana, 21 February 1801, in Karl Bruhns, *Alexander von Humboldt: Eine wissenschaftliche Biographie,* 2 vols. (Leipzig, 1872), 2: 335–44; Academia Colombiana de Ciencias Exactas, Físicas y Naturales and Academia de Ciencias de la República Democrática Alemana, eds., *Alexander von Humboldt en Columbia: Extractos de sus Diarios* (Bogota: Publicismo y Ediciones, 1982), 27–28.

5. *Alexander von Humboldt en Columbia,* 28.

6. Humboldt to Wilhelm von Humboldt, Cumana (New Granada), July 1799, in Ulrike Moheit, ed., *Briefe aus Amerika* (Berlin: Akademie Verlag, 1993), 41.

7. Humboldt to Karl Ludwig Willdenow, Havana, 21 February 1801, in Bruhns, *Alexander von Humbold,* 2:335–44.

8. *Alexander von Humboldt en Columbia,* 17.

9. Ibid.

10. Frank Baasner, *Der Begriff "sensibilité" im 18. Jahrhundert: Aufstieg und Niedergang einer Ideals* (Heidelberg: C. Winter, 1988); David Denby, *Sentimental Narrative and the Social Order in France, 1760–1820* (Cambridge: Cambridge University Press, 1994); William Reddy, "Sentimentalism and Its Erasure: The Role of the Emotions in the Era of the French Revolution," *Journal of Modern History* 72 (2000): 109–52.

11. See Lange and Jahn, eds., *Jugendbriefe,* esp. letters to Wegener, Freiesleben, and Willdenow; and Ottmar Ette, "'Eine Gemütsverfassung von moralischer Unruhe': Alexander von Humboldt und das Schreiben in der Moderne," in *Aufbruch in die Moderne* (Berlin: Akademie Verlag, 2001), 33–55. On Campe as pedagogue, children's author, and *Hauslehrer* at Tegel, see Bruhns, *Alexander von Humboldt,* 1; and essays by

Michael Niedermeier, Ulrich Herrmann, and Hans Heino Ewers in *Visionäre Lebens-klugheit: Joachim Heinrich Campe in seiner Zeit (1746–1818),* ed. H. Shmitt, Ausstellungskataloge der Herzog August Bibliothek, 74 (Wiesbaden: Harrasowitz, 1996).

12. Sergio Moravia, "Philosophie et médecine en France à la fin du XVIIIe siècle," *Studies in Voltaire and the Eighteenth Century* 39 (1972): 1089–1151; Roy Porter, "Medicine and the Science of Man in the Eighteenth Century," in *Inventing Human Science,* ed. Christopher Fox and Robert Wokler (Berkeley: University of California Press, 1995).

13. Simon Schaffer, "Natural Philosophy and Public Spectacle in the Eighteenth Century," *History of Science* 21 (1983): 1–43, and "Self-Evidence." *Critical Inquiry* 18 (1992): 327–62. On Hogarth, see Terry Castle, "The Female Thermometer," in *The Female Thermometer: Eighteenth Century Culture and the Invention of the Uncanny* (Oxford: Oxford University Press, 1995), 21–43.

14. Alexander von Humboldt, "Etwas über die lebendie Muskelfaser als anthraco-scopische Substanz," *Chemische Annalen* 2 (1795): 3–5.

15. Alexander von Humboldt, *Versuche über die gereizte Muskel und Nervenfaser, nebst Vermuthungen über den chemischen Process des Lebens in der Thier- und Pflanzenwelt,* 2 vols. (Berlin and Posen: Rottman, 1797), 1:34.

16. Ibid., 1:306.

17. Humboldt to Freiesleben, Goldmühle bei Kronach, 10 June 1793, in *Jugend-briefe,* ed. Lange and Jahn, 251; Humboldt to Freiesleben, Steben, 19 July 1793, in *Jugendbriefe,* ed. Lange and Jahn, 257–58.

18. The exact correspondence between the mines and the pass is evident in Humboldt's critique of the use of human labor to carry rocks and men in the Mexican mines (Alexander von Humboldt, *Essai politique sur le Royaume de la Nouvelle-Espagne* [Paris: J. Stone, 1811], 72–76).

19. This historical perspective is clearly conveyed in Wilhelm von Humboldt's introduction to Alexander's treatise on subterranean meteorology:

> This work will also interest those who are foreign to the technical aspects of mining and who do not busy themselves with the more exact experiments of analytic chemistry. The first part, which contains the outlines of a subterranean meteorology, reveals an almost unknown part of nature, leads into an altogether new, subterranean creation, amazes with attractive comparisons of the upper and lower atmospheres, and offers rich food not just for reflection and scientific curiosity, but for the imagination. The second part familiarizes us with the labors and dangers of mining, and if it is already in general an elevating spectacle to see the ingenuity of Man struggling with superior elements, this work awakes philanthropic sympathy for an industrious and honorable class of human beings" (in Alexander von Humboldt, *Ueber die unterirdischen Gasarten, und die Mitteln, ihren Nachtheilen zu vermindern: Ein Beytrag zur Physik der praktischen Bergbaukunde* [Braunschweig: Schulbuchhandlung, 1799], v)

20. Humboldt concludes his description of the rock's magnetism with a quotation from Condorcet's *Sketch of the Progress of the Human Mind* that insists on abstaining from theory and remaining with experience. This programmatic sensationism, empiricism, or pragmaticism runs through Humboldt's physiological, geological, and economic

work. It expresses, on the level of scientific method or epistemology, the conviction that nature is essentially dynamic and cannot be confined to human language or categories—that is, the same humanitarian sensibility that is at the center of his description of the "magnetic" mountain.

CHAPTER 4

1. Burchell to W. Hooker, 8 July 1826, Royal Botanic Gardens Kew Archives, London (hereafter cited as RBG Kew Archives).

2. Burchell to W. Hooker, 1 November 1830, RBG Kew Archives, London.

3. David Mackay, "Agents of Empire: The Banksian Collectors and Evaluation of New Lands," in *Visions of Empire: Voyages, Botany, and Representations of Nature,* ed. D. P. Miller and P. H. Reill (Cambridge: Cambridge University Press, 1996), 54.

4. David Arnold, *The Problem of Nature: Environment, Culture and European Expansion* (Oxford: Blackwell, 1996), 165–66; Richard H. Grove, *Green Imperialism: Colonial Expansion, Tropical Island Edens and the Origins of Environmentalism, 1600–1860* (Cambridge: Cambridge University Press, 1995), 73–80.

5. Grove, *Green Imperialism,* 76; Raymond Phineas Stearns, *Science in the British Colonies of America* (Chicago and London: University of Illinois Press, 1970), 44–48; David N. Livingstone, *The Geographical Tradition: Episodes in the History of a Contested Enterprise* (Oxford: Blackwell, 1992), 68–69.

6. Martin Rix, *The Art of the Botanist* (Guildford and London: Lutterworth Press, 1981), 56.

7. Grove, *Green Imperialism,* 78.

8. Rix, *The Art of the Botanist,* 79.

9. Rebecca Preston, "'The Scenery of the Torrid Zone': Imagined Travels and the Culture of Exotics in Nineteenth-Century British Gardens," in *Imperial Cities: Landscape, Display and Identities,* ed. Felix Driver and David Gilbert (Manchester: Manchester University Press, 1999).

10. For illuminating essays on Catesby's work, see A. R. W. Meyers and M. B. Pritchard, eds., *Empire's Nature: Mark Catesby's New World Vision* (Chapel Hill and London: University of North Carolina Press, 1998).

11. Ronald King, ed., *The Temple of Flora by Robert Thornton* (London: Weidenfeld & Nicolson, 1981), 13. See also Janet Browne, "Botany in the Boudoir and Garden: The Banksian Context," in *Visions of Empire,* ed. Miller and Reill, 160.

12. Nicholas Thomas, introduction to *Double Vision: Art Histories and Colonial Histories in the Pacific,* ed. Nicholas Thomas and Diane Losche (Cambridge: Cambridge University Press, 1999), 2–3.

13. See Anthony Pagden, *European Encounters with the New World* (New Haven, CT, and London: Yale University Press, 1993), 183–88.

14. Henrika Kuklick and Robert E. Kohler, eds., *Science in the Field* (Chicago: University of Chicago Press, 1996).

15. Gillian Beer, "Travelling the Other Way: Travel Narratives and Truth Claims," in *Open Fields: Science in Cultural Encounter* (Oxford: Clarendon Press, 1996), 9.

16. Luciana Martins, *O Rio de Janeiro dos Viajantes: O Olhar Britânico, 1800–1850* (Rio de Janeiro: Jorge Zahar, 2001).

17. Banks to William Townsend Aiton, 7 June 1814, RBG Kew Archives, London.

18. Cited by Kenneth Lemmon, *The Golden Age of Plant Hunters* (London: Phoenix House, 1968), 120.

19. Banks to Cunningham and Bowie, 18 September 1814, RBG Kew Archives, London.

20. Journal of the travel of Cunningham and Bowie, entry for 19 January 1815, RBG Kew Archives, London.

21. Banks from the collectors, 12 February 1815, RBG Kew Archives, London.

22. The trope of "virgin land" also colored Henry Walter Bates's impressions of the Amazonian forest in the 1850s: see Hugh Raffles, "The Uses of Butterflies: Bates of the Amazons," in *In Amazonia: A Natural History* (Princeton, NJ: Princeton University Press, 2002).

23. Cunningham to Aiton, 19 December 1818, as quoted in W. G. McMinn, *Allan Cunningham: Botanist and Explorer* (Victoria: Melbourne University Press, 1970), 10.

24. Lemmon, *The Golden Age,* 119.

25. "Instructions to Mr John Forbes employed by the Horticultural Society of London, on a mission in His Majesty's Ship Leven, commanded by Captain William Owen, to the East Coast of Africa, and other places, in the year 1822, and following years," 3 January 1822, Forbes's papers, Royal Horticultural Society Archives, London (hereafter cited as RHS Archives).

26. "Additional Instructions to Mr John Forbes. . .," 18 January 1822, Forbes's papers, RHS Archives, London.

27. The library consisted of Karl Ludwig Willdenow's *Species Plantarum* (the 10th vol. only), C. H. Persoon's *Synopsis Plantarum,* Aubert Du Petit-Thouars's *Melanges de botanique et de voyages,* João de Loureiro *Flora Cochinchinensis* (2 vols.), William Roxburgh's *Hortus Bengalensis,* Robert Sweet's *Hortus Suburbanus,* William Bullock's on preserving birds, etc., Antoine Laurent de Jussieu's *Genera Plantarum,* Thomas Bowdich's *Taxidermy,* George Samouelle's *Entomologist's Useful Compendium,* and William Turton's edition of Linnaeus's *A General System of Nature,* 7 vols.

28. The journal of Lisbon was to be sent from Lisbon; the one of Madeira, Teneriffe, and Cape Verde, from Rio de Janeiro; the one of Rio, from the Cape; and the journal of the Cape, from the Cape, and so on.

29. Forbes had difficulty finding room for his equipment and collections on the vessel and finding suitable conditions to preserve his specimens (Forbes to Joseph Sabine, Rio de Janeiro 8 June 1822, RHS Archives, London). More generally, see Nigel Rigby, "The Politics and Pragmatics of Seaborne Plant Transportation, 1769 – 1805," in *Science and Exploration in the Pacific: European Voyages in the Southern Oceans in the Eighteenth Century,* ed. Margarette Lincoln (Woodbridge: Boydell Press and the National Maritime Museum, 1998).

30. Forbes's journal, 3 May 1822, p. 28, RHS Archives, London.

31. Langsdorff's expedition to Brazil, sponsored by Tsar Alexander I, lasted seven years (1822 – 29). Among the members of the expedition were Eduard Ménétries (zoologist), Georg Freyriss (naturalist), Nester Rubtsov (astronomer), Moritzs Rugendas (artist), Ludwig Riedel (botanist), Aimé-Adrien Taunay (artist), and Hercule Florence (naval officer). See Hildegard Fauser, "O Barão Georg Heinrich von Langsdorff," in *O Brasil de Hoje no Espelho do Século XIX,* ed. Maria de Fátima, G. Costa, Pablo

Diener, and Dieter Strauss (São Paulo: Estação Liberdade, 1995), 31–34; and Ana Maria de Moraes Belluzzo, ed., *The Voyagers' Brazil,* 3 vols. (São Paulo: Metalivros, 1995), 2:124–37.

32. Forbes to Sabine, 3 June 1822, Forbes's papers, p. 15, RHS Archives, London.

33. Forbes's journal, 8 July 1822, p. 54, RHS Archives, London.

34. Sabine to Forbes, 2 May 1822, Forbes's papers, pp. 13, 14, RHS Archives, London.

35. Antonio Jozi, "Narrative of the Proceedings of a Small Party, sent by Captain Owen, of HMS Leven, in July 1823, under the Command of Lieut. C.W. Browne RN, to explore the course of the River Zambezi, on the Eastern Coast of Africa," LBR MSS 407, Royal Geographical Society Archives, London.

36. Letter from Forbes to Sabine, Algoa Bay, 29 June 1823, Forbes's papers, p. 42, RHS Archives, London.

37. Owen's report, quoted in a letter from Sabine to Reverend John Ramsay, 16 March 1824, Forbes's papers, p. 50, RHS Archives, London.

38. The practice of sketching was a valued accomplishment among naturalists, and while Burchell excelled at the art, it was also required of more humble collectors and navigators. Among Forbes's papers, for example, there is a list of drawings, referring to four plates (which have not been located): (1) Cape de Ver [*sic*] Islands; (2) Granite Rocks entrance into boto-fogo [*sic*] bay Rio Janeiro; (3) marshey [*sic*] thick woods rich soil; (4) Granite Rock boto-fogo. On the significance of sketching for navigators, see Felix Driver and Luciana Martins, "John Septimus Roe and the Art of Navigation," *History Workshop Journal* 54 (2002): 144–61.

39. Anne L. Larsen, "Not since Noah: The English Scientific Zoologists and the Craft of Collecting, 1800–1840" (Ph.D. diss., Princeton University, 1993), 37–39. See also Felix Driver, "Hints to Travellers: Observation in the Field," in *Geography Militant: Cultures of Exploration and Empire* (Oxford: Blackwell, 2001).

40. Larsen, "Not since Noah," 198.

41. Rix, *The Art of the Botanist,* 82. When William Hooker was publishing his work on exotic botany, he had difficulties in finding a sufficiently skilled engraver in Scotland to accomplish the task (Hooker to Swainson, 25 November 1821, Linnean Society Archives, London).

42. Beryl Hartley, "The Living Academies of Nature: Scientific Experiment in Learning and Communicating the New Skills of Early Nineteenth-Century Landscape Painting," *Studies in the History and Philosophy of Science* 27, no. 2 (1996): 149–80; Charlotte Klonk, *Science and the Perception of Nature: British Landscape Art in the Late Eighteenth and Early Nineteenth Centuries* (New Haven, CT, and London: Yale University Press, 1996); Luciana Martins, "The Art of Tropical Travel, 1768–1830," in *Georgian Geographies,* ed. Miles Ogborn and Charles Withers (Manchester: Manchester University Press, 2004).

43. Raffles, "The Uses of Butterflies," 130–31.

44. Alexander von Humboldt, *Voyages aux régions équinoxiales du nouveau continent,* 30 vols. (Paris; 1814–25), and *Personal Narrative of Travels to the Equatorial Regions of the New Continent,* 7 vols. (London, 1814–29).

45. Alexander von Humboldt, *Personal Narrative of a Journey to the Equinoctial Regions of the New Continent* (London: Penguin Books, 1995), 297.

46. Anne Godlewska, "From Enlightenment Vision to Modern Science? Humboldt's Visual Thinking," in *Geography and Enlightenment,* ed. David N. Livingstone and Charles W. J. Withers (Chicago: University of Chicago Press, 1999), 267.

47. Michael Dettelbach, "The Face of Nature: Precise Measurement, Mapping and Sensibility in the Work of Alexander von Humboldt," *Studies in History and Philosophy of Biological and Biomedical Sciences* 30, no. 4 (1999): 473−504.

48. Helen M. McKay, introduction to *The South African Drawings of William Burchell,* ed. Helen M. McKay, 2 vols. (Johannesburg: Witwatersrand University Press, 1952), 2:xv.

49. Klonk, *Science and the Perception of Nature,* 80.

50. For a study of Burchell's environmental work in St. Helena, see Grove, *Green Imperialism,* 343−64.

51. William John Burchell, n.d., "St. Helena Plants," RBG Kew Archives, London. On natural history drawings as proxy specimens, see Martin Rudwick, "Georges Cuvier's Paper Museum of Fossil Bones," *Archives of Natural History* 27 (2000): 51−68.

52. William John Burchell, *Travels in the Interior of Southern Africa,* 2 vols. (London: Batchworth Press, 1953), 2:152.

53. Edward B. Poulton, *William John Burchell* (London: Spottiswoode, 1907), 39−40.

54. Driver, *Geography Militant,* 18−19.

55. Burchell, *Travels,* 2:194, 371.

56. Ibid., 396.

57. Swainson arrived in the province of Pernambuco in December 1816. Although he had planned a journey deep into the interior, political unrest confined his research to a rather limited area. When the rebellion had been quelled, Swainson left Pernambuco and made for the river St. Francisco. He continued his journey to Bahia and then embarked for Rio. He returned to England in August 1818. Swainson was very critical of "the artificial habits and the luxury of English society". In a letter to Audubon, dated March 1830, he vented his feelings: "I am sick of the world and of mankind, and but for my family would end my days in my beloved forests of Brazil" (Geoffrey M. Swainson, ed., *William Swainson, Naturalist and Artist* [Palmerston North: Swift, 1992], 30).

58. A collection of drawings and watercolors by Landseer, Burchell, Chamberlain, and Debret executed during Stuart's mission is now held by Instituto Moreira Salles in Rio de Janeiro; see *Brasil, 1825−26: Charles Landseer e a Missão Britânica e Trabalhos de Burchell, Chamberlain, Debret,* Christie's auction's catalog (London: Christie's, 1997).

59. Burchell to Salisbury, 5 September 1826, Linnean Society Archives, London.

60. Burchell to William Hooker, 8 July 1816, RBG Kew Archives, London.

61. Maria Graham's first voyage to Brazil was from 1821 to 1823 and the second, from 1824 to 1825. In 1824, preparing to sail back to Rio, Maria Graham offered her services to William Hooker: "I do not habitually draw flowers but I could do that— and also any peculiar form of seed &c—Only let me know how I can be useful & I will try to be so" (Maria Graham to William Hooker, 11 April 1824, RBG Kew Archives, London). A portfolio with a hundred drawings she made during her stay in Rio also survives in the RBG Kew Archives.

62. Hamond described Graham as a "he-she" (Paulo F. Geyer, ed., *Os Diários do Almirante Graham Eden Hamond: 1825–1834/38* [Rio de Janeiro: Editora JB, 1984], 27).

63. Poulton, *William John Burchell,* 43−44.

64. Burchell to W. Hooker, 11 December 1830, RBG Kew Archives, London.

65. See R. F. Kennedy, comp., *Catalogue of Pictures in the Africana Museum,* vol. 6, *Supplement A–G* (Johannesburg: Africana Museum, 1971), B2131–B2351; and Gilberto Ferrez, *O Brasil do Primeiro Reinado Visto pelo Botânico William John Burchell 1825/1829* (Rio de Janeiro: Fundação João Moreira Salles/Pró-Memória, 1981).

66. Burchell to W. Hooker, 8 July 1826, RBG Kew Archives, London. The panorama is published in Gilberto Ferrez, *Rio de Janeiro's Most Beautiful Panorama by William John Burchell, 1825/1829* (Rio de Janeiro: Instituto Histórico e Geográfico Brasileiro, 1966).

67. Burchell to Mrs. Burchell, 29 August 1826, Oxford University Museum Archives, Oxford.

68. Burchell to Mrs. Burchell, 10 June 1827, Oxford University Museum Archives, Oxford.

69. Burchell to W. Hooker, 18 April 1830, RBG Kew Archives, London.

70. Burchell to Matthew Burchell, 29 November 1826, Oxford University Museum Archives, Oxford. The boy eventually returned to England as Burchell's servant (Burchell to Mrs. Burchell, 1 December 1829, Oxford University Museum Archives, Oxford). Jane Pickering has identified a reference to a black servant in Burchell's Fulham residence within a letter of 1830 by Mary Swainson ("William John Burchell's Travels in Brazil, 1825−1830, with Details of the Surviving Mammals and Bird Collections," *Archives of Natural History* 25, no. 2 [1998]: 237−66).

71. Jane R. Camerini, "Wallace in the Field," *Osiris* 11 (1996): 44−65.

72. Burchell to Salisbury, 5 September 1826, Linnean Society Archives, London; see also Burchell to W. Hooker, 8 July 1826, RBG Kew Archives, London.

73. Bruno Latour, "Drawing Things Together," in *Representation in Scientific Practice,* ed. Michael Lynch and Steve Woolgar (Cambridge, MA: MIT Press, 1990).

74. Burchell, *Travels,* 1:505.

75. Pickering, "William John Burchell's Travels in Brazil."

76. Edward B. Poulton, *The Collections of William John Burchell, D.C.L., in the Hope Department* (Oxford: Oxford University Museum, 1903), 1−12.

77. Raffles, "The Uses of Butterflies," 136.

78. David Allen, "Tastes and Crazes," in *Cultures of Natural History,* ed. Nicholas Jardine, James A. Secord, and Emma C. Spary (Cambridge: Cambridge University Press, 1996), 394.

79. Anne Secord, "Botany on a Plate: Pleasure and the Power of Pictures in Promoting Early Nineteenth-Century Scientific Knowledge," *Isis* 93, no. 1(2002): 39.

80. Kuklick, introduction, 4.

81. David Allen, *The Naturalist in Britain: A Social History* (London: Penguin Books, 1976), 101.

82. Sue Minter, *The Greatest Glass House: The Rainforests Recreated* (London: HMSO, 1990), 3.

83. Larsen, "Not since Noah," 90.

84. Darwin to R. W. Darwin, 8 February−1 March 1832, in F. H. Burkhardt and

S. Smith, eds., *The Correspondence of Charles Darwin* (Cambridge: Cambridge University Press, 1985), 1:202. See also Luciana Martins, "A Naturalist's Vision of the Tropics: Charles Darwin and the Brazilian Landscape," *Singapore Journal of Tropical Geography* 21, no. 1 (2000): 19-33.

85. Beer, *Open Fields*, 59.

86. Stephen Bann, "Travelling to Collect: The Booty of John Bargrave and Charles Waterton," in *Travellers' Tales: Narratives of Home and Displacement*, ed. George Robertson, Melinda Mash, Lisa Tickner, Jon Bird, Barry Curtis, and Tim Putman (London: Routledge, 1994), 160.

CHAPTER 5

The epigraph is from Derek Walcott's poem "Midsummer" (1984), in *Collected Poems, 1948-1984* (New York: Farrar, Straus & Giroux, 1986), 480.

1. See David Arnold, "Inventing Tropicality," in *The Problem of Nature: Environment, Culture and European Expansion* (Oxford: Blackwell, 1996); and Nancy Leys Stepan, *Picturing Tropical Nature* (London: Reaktion Books, 2001).

2. Christopher Columbus, *The "Diario" of Christopher Columbus's First Voyage to America: 1492-93,* ed. and trans. Oliver Dunn and James E. Kelley (Norman: University of Oklahoma Press, 1989), 67.

3. Peter Martyr, *De novo Orbe; or, The Historie of the West Indies,* trans. Richard Eden and Michael Lok (London, 1612), 140v. This fertility had been a feature of the Fortunate Islands, usually identified as the Canaries: "Its fields have no need of the farmer's plow. . . . Its soil bears everthing as if it were grass, by spontaneous production" (Geoffrey of Monmouth's *Vita Merlini,* quoted in George Boas, *Essays on Primitivism and Related Ideas in the Middle Ages* [1948; reprint, New York: Octagon Books, 1978], 169).

4. Martyr, *De novo Orbe,* 51r.

5. See Peter Hulme, "Making Sense of the Native Caribbean," *New West Indian Guide* 67, nos. 3 and 4 (1993): 189-220.

6. Quoted in Philippe Despoix, "Naming and Exchange in the Exploration of the Pacific: On European Representations of Polynesian Culture in Late XVIII Century," in *Multiculturalism and Representation: Selected Essays,* ed. John Rieder and Larry E. Smith (Honolulu: University of Hawai'i, 1996), 4. See also L. Davis Hammond, ed., *News from New Cythera: A Report of Bougainville's Voyage, 1766-1769* (Minneapolis: University of Minnesota Press, 1970).

7. Quoted in Bernard Smith, *European Vision and the South Pacific,* 2d ed. (New Haven, CT, and London: Yale University Press, 1985), 48.

8. George Hamilton, "A Voyage Round the World," in *Voyage of H.M.S. "Pandora" . . . ,* ed. Basil Thomson (London: Francis Edwards, 1915), 108. In his *Génie du Christianisme,* Chateaubriand also wrote of how Tahitians found "le lait et le pain suspendus aux branches des arbres" (quoted in Smith, *European Vision,* 153).

9. J. R. Forster, *Observations Made on a Voyage Round the World* (London: Robinson, 1778), 228.

10. Nicholas Thomas, "The Force of Ethnology: Origins and Significance of the Melanesia/Polynesia Division," *Current Anthropology* 30 (1989): 27-41, at 31. See also Vanessa Agnew, "Pacific Island Encounters and the German Invention of Race," in

Islands in History and Representation, ed. Rod Edmond and Vanessa Smith (London: Routledge, 2003).

11. See Bronwen Douglas, "Science and the Art of Representing 'Savages': Reading 'Race' in Text and Image in South Seas Voyage Literature," *History and Anthropology* 11, nos.2−3 (1999): 157−201, at 180.

12. The idea that the world has three physical environments goes back to Claudius Ptolemy and, beyond him, to Hippocrates and Aristotle. Ptolemy has three regions (northern, southern, intermediate), with black Ethiopians to the south and white Scythians to the north. Ethiopians are "savage" because "their homes are continually oppressed by the heat." Scythians "because their dwelling places are continually cold." Those in between, Ptolemy describes as "medium in colouring, of moderate stature, in nature equable, live close together, and are civilized in their habits" (Clarence J. Glacken, *Traces on the Rhodian Shore: Nature and Culture in Western Thought from Ancient Times to the End of the Eighteenth Century* [Berkeley: University of California Press, 1967], 84, 112).

13. Glacken, *Traces,* 437−38.

14. James Kennedy, "On the Probable Origin of the American Indians, with Particular Reference to the Caribs," in *Ethnological Essays* (London: Arthur Hall, Virtue & Co., 1855).

15. Harold Sterling Gladwin, *Men Out of Asia* (New York: McGraw-Hill Book Co., 1949), 250−51.

16. Johann Friedrich Blumenbach, *Anthropological Treatises,* ed. and trans. Thomas Bendyshe (London: Longman, Green, Longman, Roberts, & Green, 1865), 156.

17. See Paul Turnbull, "Enlightenment Anthropology and the Ancestral Remains of Australian Aboriginal People," in *Voyages and Beaches: Pacific Encounters, 1769−1840,* ed. Alex Calder, Jonathan Lamb, and Bridget Orr (Honolulu: University of Hawai'i Press, 1999).

18. Blumenbach, *Anthropological Treatises,* 162.

19. Ibid., 107.

20. See Milford Wolpoff and Rachel Caspari, *Race and Human Evolution: A Fatal Attraction* (Boulder, CO: Westview Press, 1997), 62−63.

21. Jean Baptiste Du Tertre, "Histoire générale des Antilles habitées par les François" (1667), in *Wild Majesty: Encounters with Caribs from Columbus to the Present Day,* ed. Peter Hulme and Neil L. Whitehead (Oxford: Clarendon Press, 1992), 129.

22. See *Wild Majesty,* ed. Hulme and Whitehead, chaps. 14−19; Peter Hulme, *Colonial Encounters: Europe and the Native Caribbean, 1492−1797* (London: Methuen, 1986), 225−63, and "Black, Yellow, and White on St. Vincent: Moreau de Jonnès's Carib Ethnography," in *The Global Eighteenth Century,* ed. Felicity A. Nussbaum (Baltimore: Johns Hopkins University Press, 2003).

23. Bryan Edwards, *The History, Civil and Commercial, of the British Colonies in the West Indies* (1793), 2 vols. (New York: Arno Press, 1972), 1:77−78. Extensive comparison follows, 78−81.

24. The story is retold in Antonello Gerbi, *The Dispute of the New World: The History of a Polemic, 1750−1900,* trans. Jeremy Moyle (1955; reprint, Pittsburgh: University of Pittsburgh Press, 1973), 40.

25. See Peter Hulme, "The Spontaneous Hand of Nature: Savagery, Colonialism,

and the Enlightenment," in *The Enlightenment and Its Shadows,* ed. Peter Hulme and Ludmilla Jordanova (London: Routledge, 1990).

26. Joseph Banks, *Manners & Customs of S. Sea Islands* (1769), http://coombs .anu.edu.au/~cookproj/archive/gen_rem_tahiti/bt009.html (with Banks's idiosyncratic spelling). The trope is present in Pierre Loti (and beyond): "In Oceania toil is a thing unknown. The forests spontaneously produce all that is needed for the support of these unforeseeing races; the fruit of the bread-tree, and wild bananas grow for all the world to pluck, and suffice for their need. The years glide over the Tahitians in utter idleness and perpetual dreaming, and these grown-up children could never conceive that in our grand Europe there should be so many people wearing out their lives in earning their daily bread" (*The Marriage of Loti* [1880], trans. Wright Frierson and Eleanor Frierson [Honolulu: University Press of Hawai'i, 1976], 40).

27. See William Denevan, *Cultivated Landscapes of Native Amazonia and the Andes* (Oxford: Oxford University Press, 2001); and Dana Lepofsky, "Gardens of Eden? An Ethnocentric Reconstruction of Maohi (Tahitian) Cultivation," *Ethnohistory* 46, no.1 (1999): 1−29.

28. Paul Fussell, *Abroad: British Literary Traveling Between the Wars* (Oxford: Oxford University Press, 1980), 3−8.

29. Alec Waugh, *The Early Years of Alec Waugh* (London: Cassell, 1962), 214.

30. Ibid., xiii.

31. Alec Waugh, *Hot Countries* (New York: Literary Guild, 1930), 119.

32. Ibid., 73.

33. Waugh, *The Early Years,* 291.

34. Alec Waugh, *The Sugar Islands* (London: Cassell, 1958), 281, 284.

35. Alec Waugh, *The Fatal Gift* (London: W. H. Allen & Co., 1973). The phrase comes from Lord Byron, "Childe Harold's Pilgrimage," canto 4, lines 42−43, in *The Complete Poetical Works,* ed. Jerome J. McGann, 7 vols. (Oxford: Clarendon Press, 1980), 2:138−39, but Byron is translating the phrase "dono infelice di bellezza" from a sonnet by the seventeenth-century Italian writer, Vincenzo da Filicaia. On the Italian connection to these issues of tropicality, see Peter Hulme, "*The Fatal Gift:* Storie di fine impero," in *Le maschere letterare l'impero britannico,* ed. Daniela Corona and Elio di Piazza (Pisa: Edizioni ETS, 2005).

36. According to Loren Baritz, happiness and death "formed the dialectic of the west" ("The Idea of the West," *American Historical Review* 66 [1960−61]: 620). David Arnold similarly notes the deep ambivalence at the heart of the idea of tropicality ("Inventing Tropicality," 142).

CHAPTER 6

Thanks are due to Denis Cosgrove and Luciana Martins for their insightful comments on an earlier draft and to Geoff Hancock for information about the Goliath beetle.

1. The paper, presented in a letter to Sir Joseph Banks, was read to the society on 15 February 1781 (Henry Smeathman, "Some Account of the Termites, Which Are Found in Africa and Other Hot Climates," *Philosophical Transactions of the Royal Society* 71 [1781]: 139−92).

2. Maurice Polydore Marie Bernard Maeterlinck, *The Life of the White Ant,* trans. Alfred Sutro (London: George Allen & Unwin, 1927), 19.

3. Smeathman's career as a naturalist and his schemes for settlement are the subject of Starr Douglas, "Natural History, Improvement and Colonisation: Henry Smeathman and Sierra Leone in the Late Eighteenth Century" (Ph.D. diss., Royal Holloway, University of London, 2004).

4. On Sandby, see Michael Rosenthal, *British Landscape Painting* (Oxford: Phaidon, 1982); Jessica Christian, "Paul Sandby and the Military Survey of Scotland," in *Mapping the Landscape: Essays on Art and Cartography,* ed. Nicholas Alfrey and Stephen Daniels (Nottingham: Castle Museum, 1990), 18–22; Charlotte Klonk, *Science and the Perception of Nature: British Landscape Art in the Late Eighteenth and Early Nineteenth Centuries* (New Haven, CT, and London: Yale University Press, 1996).

5. Barbara Maria Stafford, *Voyage into Substance. Art, Science, Nature, and the Illustrated Travel Account, 1760–1840* (Cambridge, MA: MIT Press, 1984), 16; Bernard Smith, *European Vision and the South Pacific, 1768–1850* (London: Oxford University Press, 1960).

6. Klonk, *Science and the Perception of Nature,* 71.

7. Ibid., 72.

8. Smeathman, "Some Account," 156.

9. Ibid., 158–59.

10. Henry Smeathman, *Mémoir pour servir à l'histoire de quelques insectes, connus sous les noms de Termés, ou fourmis blanches . . . accompagné de figures gravées en taille-douce,* trans. Jean Cyrille Rigaud (Paris, 1786), 25–26.

11. Barbara Maria Stafford, "Presuming Images and Consuming Words: The Visualization of Knowledge from the Enlightenment to Post-modernism," in *Consumption and the World of Goods,* ed. John Brewer and Roy Porter (London: Routledge, 1993), 468. See also Barbara Maria Stafford, *Body Criticism: Imagining the Unseen in Enlightenment Art and Medicine* (Cambridge, MA: MIT Press, 1993).

12. Smeathman ("Some Account," 172n29) refers to the dissection of some of his specimens by John Hunter in support of one of his calculations concerning the rate of production of eggs by termite queens.

13. Susan Dixon, "The Sources and Fortunes of Piranesi's Archaeological Illustrations," in *Tracing Architecture: The Aesthetics of Antiquarianism,* ed. Dana Arnold and Stephen Bending (Oxford: Blackwell, 2003), 49.

14. Smeathman, "Some Account," 140.

15. Ibid., 139.

16. Our approach in this section reflects the argument of Kay Dian Kriz in "Curiosities, Commodities, and Transplanted Bodies in Hans Sloane's 'Natural History of Jamaica,'" *William and Mary Quarterly* 57 (2000): 35–78.

17. "When these hills are at about little more than half their height, it is always the practice of the wild bulls to stand as centinels [*sic*] upon them, while the rest of the herd is ruminating below" (Smeathman, "Some Account," 151).

18. Timothy Clayton, *The English Print, 1688–1802* (New Haven, CT, and London: Yale University Press, 1997), 176; and Rüdiger Joppien and Bernard Smith, eds., *The Art of Captain Cook's Voyages,* 2 vols. (New Haven, CT, and London: Yale University Press, 1985), 2:226.

19. Smith, *European Vision,* 9.

20. Ibid.; Klonk, *Science and the Perception of Nature,* 67−76.

21. Martin Rudwick, "The Emergence of a Visual Language for Geological Science, 1760−1840," *History of Science* 14 (1976): 155.

22. See also Beryl Hartley, "The Living Academies of Nature: Scientific Experiment in Learning and Communicating the New Skills of Early Nineteenth-Century Landscape Painting," *Studies in the History and Philosophy of Science* 27 (1996): 149−80.

23. A full account of Smeathman's fieldwork in Sierra Leone and the Caribbean is provided in Douglas, "Natural History, Improvement and Colonisation," chap. 4.

24. Bernard Smith famously uses the case of Sherwin's 1777 engraving, after William Hodges, of *The Landing at Middleburgh* as a paradigm for the way in which European classicism was imposed on a more descriptive view of Pacific peoples; see also chap. 2, this volume.

25. Smeathman, "Some Account," 151.

26. Nancy Leys Stepan has noted how the close foregrounding of objects "gives an intensified sense of the alterity of nature in the tropics, or a sense of its enormity or strangeness"—this, though, was only one side of the story (*Picturing Tropical Nature* [London: Reaktion Books, 2001], 69).

27. Stafford, *Voyage into Substance,* 41.

28. Kriz, "Curiosities," 77.

29. On the Banksian empire, see esp. David Miller and Peter Reill, eds., *Visions of Empire: Voyages, Botany and Representations of Nature* (Cambridge: Cambridge University Press, 1996); Richard Drayton, *Nature's Government: Science, Imperial Britain and the "Improvement" of the World* (New Haven, CT, and London: Yale University Press, 2000).

30. Smeathman's sponsors also included Marmaduke Tunstall and the duchess of Portland (Douglas, "Natural History, Improvement and Colonisation," 63−80).

31. Dru Drury, *Illustrations of Natural History: Wherein Are Exhibited Upwards of Two Hundred and Forty Figures of Exotic Insects, according to their Different Genera; . . . with a Particular Description of Each Insect,* 3 vols. (London, 1770−82), 1:67.

32. In the catalog to the Hunter Museum in Glasgow, where the specimen is kept, it is still classified as *Goliathus goliatus* (Drury); see Robert Staig, *The Fabrician Types of Insects in the Hunterian Collection at Glasgow University: Coleoptera,* pt. 1 (Cambridge: Cambridge University Press, 1931), 86.

33. C. H. Brock, "Dru Drury's *Illustrations of Natural History* and the Type Specimen of *Goliathus goliatus* Drury," *Journal of the Society for the Bibliography of Natural History* 8, no. 3 (1977): 259.

34. Dru Drury to Dr. Pallas, 28 February 1768, in T. Cockerell, "Dru Drury, an Eighteenth-Century Entomologist," *Scientific Monthly* (January 1922), 71.

35. F. W. Hope, "The Auto-Biography of John Christian Fabricius," *Abstractions of the Entomological Society* 4 (1845−47): v−vi. Fabricius visited London again in 1772−75, when he was a close associate of Banks, Hunter, and Drury, and in 1780, when he visited Smeathman, among others.

36. Brock, "Dru Drury's *Illustrations,*" 259−63.

37. Anne Larsen, "Not Since Noah: The English Scientific Zoologists and the Craft of Collecting, 1800−1840 (Ph.D. diss., Princeton University, 1993), 234.

38. Dru Drury, Letter-book of Dru Drury, 1761–1803, Natural History Museum, Entomological Library, London, SB FD6, 86; Bill Noblett, "Dru Drury's 'Directions for Collecting Insects in Foreign Countries,'" *Amateur Entomologist's Society Bulletin* 44 (1985): 170–78.

39. Henry Smeathman, Account of the Tarantula of Sierra Leone (n.d.), Linnean Society, London, MSS 271. Harris's illustration is filed together with Smeathman's description, which was read to the society in November 1802.

40. P. E. H. Hair, Adam Jones, and Robin Law, eds., *Barbot on Guinea. The Writings of Jean Barbot on West Africa, 1678–1712* (London: Hakluyt Society, 1992), 469–78.

41. William Smith, *A New Voyage to Guinea: Describing . . . Whatever . . . Is Memorable among the Inhabitants. Likewise an Account of the Animals, Minerals, Etc.* (London: J. Nourse, 1744), 150–54.

42. Drury also devoted a substantial essay in the third volume to Smeathman's collecting activities in Sierra Leone, a large part of which was devoted to problems of preservation and transportation in tropical regions (Drury, *Illustrations of Natural History,* 3:v–xxvi).

43. John Coakley Lettsom, ed., *The Works of J. F.* [John Fothergill] (London, 1784), 575.

44. Smeathman to Drury, 15 January 1774, Mr. Smeathman's letters to Mr. Drury, Ms. D. 26, Uppsala University Library, Uppsala, 37.

45. Smeathman to Lettsom, 19 October 1782, in Lettsom, *Works of J. F.,* 580–81.

46. Smeathman's critical views on indigenous rice cultivation reflected his assumptions about the superiority of European techniques of harvesting. See also Judith Carney, *Black Rice: The African Origins of Rice Cultivation in the Americas* (Cambridge, MA, and London: Harvard University Press, 2001).

47. Henry Smeathman, *Plan of Settlement to Be Made Near Sierra Leona, on the Grain Coast of Africa* (London, 1786).

48. "Henry Smeathman to Dr. Knowles, 21 July 1773," *New Jerusalem Magazine* (1790), 280.

49. Ibid., 279. See also David Mackay, "Agents of Empire: The Banksian Collectors and Evaluation of New Lands," in *Visions of Empire,* ed. Miller and Reill.

50. "Henry Smeathman to Dr. Knowles," 289–90.

51. Ibid., 283.

52. Thomas Masterman Winterbottom, *An Account of the Native Africans in the Neighbourhood of Sierra Leone,* 2 vols. (London: John Hatchard, 1803), 1:292–336.

53. See "*Smeathmannia laevigata* (Smooth-Stalked Smeathmannia)," *Curtis's Botanical Magazine* 71 (1845), plate 4194; and "*Smeathmannia pubescens* (Downy Smeathmannia)," *Curtis's Botanical Magazine* 74 (1848), plate 4364. On *Pyralis Smeathmanniana,* see Johann Christian Fabricius, *Species insectorum, exhibentes eorum differentias specificas, synonima auctorum, loca natalia, metamorphosin: adjectis observationibus, descriptionibus,* 3 vols. (Hamburg and Kiel, 1781), 3:278.

54. N. M. Collins, "Two Hundred Years of Termitology," *Antenna: Bulletin for the Royal Entomological Society* 4, no. 2 (1980): 42–48.

55. For recent versions of Smeathman's cross sections, see P. J. Gullan and P. S. Cranston, *The Insects: An Outline of Entomology* (London: Chapman & Hall, 1994), 316; P. E. Howse, *Termites: A Study in Social Behaviour* (London: Hutchinson Univer-

sity Library, 1970), 107; Edward Osborne Wilson, *The Insect Societies* (Cambridge, MA: Harvard University Press, Belknap Press, 1971), 119.

CHAPTER 7

This work has benefited from constructive discussions with the fellows of the Old Dominion Group at Princeton University and the Pacific History Reading Group at MIT/Harvard. I would also like to thank Katherine Anderson, Michael Bravo, Christina Burnett, Denis Cosgrove, Angela Creager, Steven Dick, Felix Driver, Ronald Grim, Michael Mahoney, Helen Rozwadowski, James Secord, Christine Stansell, and the participants in the "Tropical Views and Visions" conference at the National Maritime Museum, Greenwich, July 2002.

1. Note that Melville makes dramatic symbolic use of navigational technologies elsewhere in *Moby-Dick* (as well as in his other writings); see, saliently, chap. 118, "The Quadrant," where Ahab destroys his equipment for "celestial" orientation. There is a very large literature on these matters, which I will not rehearse here.

2. Nota bene: this was not the actual map Maury was promising (which would come in four sheets over the next two years), but this preliminary map was meant as a teaser for the complete series of whale charts that were in the process of engraving. This was a common practice in the map trade, a way of both gauging and stimulating interest in a new product. Where did Melville see the circular? One good possibility is *Hunt's Merchants' Magazine and Commercial Review,* where the full circular appeared in the June issue. But the announcement was also reported in several New York newspapers. For some of this, see Howard P. Vincent, *The Trying-out of Moby-Dick* (New York: Houghton Mifflin, 1949). For a brief discussion of the whale charts in the context of the sperm whale fishery in the Pacific, see chap. 8 of Granville Allen Mawer, *Ahab's Trade: The Saga of South Sea Whaling* (New York: St. Martins, 1999), 258–74. A number of earlier whaling historians (and some whale biologists) have made use of the whale charts to reconstruct the movements and distributions of both whales and whale ships in the nineteenth century. I know of only one recent essay that tries (not entirely successfully, in my view) to read Maury in connection with *Moby-Dick:* T. Hugh Crawford, "Networking the (Non) Human: *Moby-Dick,* Matthew Fontaine Maury, and Bruno Latour," *Configurations* 5, no. 1 (1997):1–21.

3. The most exhaustive biographical work is Francis Leigh Williams, *Matthew Fontaine Maury, Scientist of the Sea* (New Brunswick, NJ: Rutgers University Press, 1963). See also Charles Lee Lewis, *Matthew Fontaine Maury: The Pathfinder of the Seas* (Annapolis, MD: U.S. Naval Institute, 1927); Columbus O. Iselin, *Matthew Fontaine Maury (1806–1873), Pathfinder of the Seas: The Development of Oceanography* (New York: Newcomen Society, 1957); Patricia Jahns, *Matthew Fontaine Maury and Joseph Henry: Scientists of the Civil War* (New York: Hastings House, 1961); and John Walter Wayland, *The Pathfinder of the Seas: The Life of Matthew Fontaine Maury* (Richmond, VA: Garrett & Massie, 1930). A somewhat hagiographic program attends much of this work and is not entirely avoided by more recent contributions, for example, Andreas B. Rechnitzer, "Matthew Fontaine Maury and Friends?" in *Sky with Ocean Joined: Proceedings of the Sesquicentennial Symposia of the U.S. Naval Observatory,* ed. Steven Dick and LeRoy E. Doggett (Washington, DC: U.S. Naval Observatory, 1983); and Chester

G. Hearn, *Tracks in the Sea: Matthew Fontaine Maury and the Mapping of the Oceans* (Camden, ME: International Marine, 2002). For a critical analysis (discussed below), see Harold Burstyn's entry in the *Dictionary of Scientific Biography* (s.v.) or John Leighly's contribution to the published communications of the Premier Congrès International d'Histoire de l'Océanographie, "M. F. Maury in His Time," *Bulletin de l'Institut Océanographique,* Fondation Albert I, Prince de Monaco, special no. 2 (1968): 147−61. Another significant deflationary account can be found in Leighly's introduction to Maury's *The Physical Geography of the Sea,* ed. John Leighly (Cambridge, MA: Harvard University Press, 1963). See also the essay review by Nathan Reingold, "Two Views of Maury . . . and a Third," *Isis* 55, no 3 (1964): 370−72. I originally drafted this chapter in 2002, without benefit of Steven Dick's excellent discussion of Maury's work at the Naval Observatory, "A Choice of Roles: The Maury Years, 1844−61," chap. 2 of *Sky and Ocean Joined: The U.S. Naval Observatory, 1830−2000* (Cambridge: Cambridge University Press, 2003), 60−117, but his detailed and balanced assessment should be a point of departure for future Maury work.

4. For a detailed discussion of editions, see Williams, *Matthew Fontaine Maury,* 698−99. These figures must be treated with some caution; most of the American editions were reprints, and while there was a "19th edition" printed in London in 1883, there is no evidence of many of the lower-numbered editions. It was not uncommon for publishers to slap a high number on such a volume to communicate that it was a must-have book.

5. The U.S. Naval Observatory was created by an act of the U.S. Congress in 1842 and realized in stone on the muggy fens of the capital in 1844. On this institution, see Dick, *Sky and Ocean Joined;* Dick and Doggett, eds., *Sky with Ocean Joined;* and Steven J Dick, "John Quincy Adams, the Smithsonian Bequest and the Founding of the U.S. Naval Observatory," *Journal for the History of Astronomy* 22 (1991): 29−44, and, regarding the forerunner establishment, "Centralizing Navigational Technology in America: The U.S. Navy's Depot of Charts and Instruments, 1830−1842," *Technology and Culture* 33 (1992): 467−509.

6. Maury's treatise was first published in 1836, as *A New Theoretical and Practical Treatise on Navigation; together with a new and easy plan for finding diff. lat., course, and distance, in which the auxiliary branches of mathematics and astronomy—comprising algebra, geometry, variation of the compass, etc.—are treated. Also, the theory and most simple method of finding time, latitude, and longitude.* It was revised in 1843 as *Elementary, Practical and Theoretical Treatise on Navigation.* Also sometimes called *Maury's Navigation,* it replaced Nathaniel Bowditch's *New American Practical Navigator,* which itself had replaced John Hamilton Moore's *Practical Navigator.*

7. Two of them appeared in Benjamin Silliman's *American Journal of Science and Arts* in 1834 (see Williams, *Matthew Fontaine Maury,* 693).

8. See Helen Rozwadowski, "Fathoming the Ocean: Discovery and Exploration of the Deep Sea, 1840−1880" (Ph.D. diss., University of Pennsylvania, 1996); Roger H. Charlier and Patricia S. Charlier, "Matthew Fontaine Maury, Cyrus Field and the Physical Geography of the Sea," *Sea Frontiers* 16, no. 5 (1970): 272−81.

9. On Maury in the history of cartography, see Robert C. Hansen, "The Cartographic Contributions of Matthew Fontaine Maury" (n.d.), typescript, Library of Congress, Geography and Map Divisions, Pamphlet File; Herman R. Friis, "High-

lights of Matthew Fontaine Maury's Life as a Virginia Geographer in the Service of His Country and his State," *Virginia Geographer* 20 (1988): 27−32; and *United States Geographical Exploration of the Pacific Basin, 1783−1899,* National Archives publication no. 62-2 (Washington, DC: National Archives, National Archives and Records Service, General Services Administration, 1961).

10. This quote from Maury is found in the much-expanded eighth edition of *The Physical Geography of the Sea,* published in 1861, the edition from which Leighly prepared the Harvard University Press reprint edition of 1963 (389); wherever possible I will give page citations to this edition below, but comparative work with the earlier editions is sometimes necessary and often illuminating. There is as yet no variorum edition of *The Physical Geography of the Sea.* Katherine Anderson points out that Maury's success with the maritime observation scheme must be understood in the context of the considerably slower contemporary progress toward international cooperation in land-based meteorological data collection. See her forthcoming book, *Predicting the Weather: Victorians and the Science of Meteorology* (Chicago: University of Chicago Press, 2005).

11. Maury, *The Physical Geography of the Sea,* ed. Leighly, 6.

12. Hearn, *Tracks in the Sea,* 246. Note that for the majority of this period, Maury would no longer have been the head of the observatory.

13. Several foreign sources do confirm interest by Britain and France; see "Ocean Currents: Remarks on Maury's Wind and Current Charts, as Alluded to in Lord Wrottesley's Speech, in the House of Lords, on the 25th of April, 1853," *Nautical Magazine* 22 (1853): 475−549; "Documents Nautiques," *Revue Coloniale,* ser. 2, 17 (1857): 419−51 and plates.

14. Maury received gifts from the "Merchants and Underwriters of New York" in 1853; see Williams, *Matthew Fontaine Maury,* 194. For his links to the Board of Underwriters, see 266. Further evidence for the utility of the charts lies in the successful calls for their reissue after a Civil War hiatus. The point is made by Marc I. Pinsel in his useful article, "The Wind and Current Chart Series Produced by Matthew Fontaine Maury," *Navigation* 28 (1981): 123−36, esp. 137.

15. These issues have now been discussed in very helpful detail by Dick, *Sky and Ocean Joined,* 98−117, who is, on the whole, more sympathetic to Maury's standing as an administrator of astronomical work than an earlier generation of commentators in the history of science.

16. It is worth bearing in mind that, as Dick shows, at least part of what is at issue here is a dispute over the proper function of an "observatory" in this period. Were these institutions primarily for the purpose of navigational astronomy? More than a century of observatory labor at places like Greenwich had been dedicated to archiving lunar movement and stellar positions for the purpose of realizing an astronomical solution to the longitude problem. See William J. H. Andrewes, *Quest for Longitude* (Cambridge, MA: Collection of Historical Instruments at Harvard University, 1996); and Derek Howse, *Greenwich Time and the Discovery of the Longitude* (Oxford: Oxford University Press, 1979).

17. Note that Maury was himself supposed to serve in the expedition as its astronomer but withdrew when Wilkes took the command. The machinations that attended the expedition have been worked through by William Stanton, *The Great*

United States Exploring Expedition of 1838–1841 (Berkeley: University of California Press, 1975). For a more recent review of some of the issues, see Herman Viola and Carolyn Margolis, eds., *Magnificent Voyagers: The United States Exploring Expedition 1838–1842* (Washington, DC: Smithsonian Press, 1985); and Nathaniel Philbrick, *Sea of Glory* (New York: Viking, 2003).

18. Williams, *Matthew Fontaine Maury,* 364.

19. On his return he taught at the Virginia Military Institute.

20. The precedents were largely meteorological, including, for example, the work of the Connecticut harness maker William C. Redfield, who in 1831 and 1833 published original pieces tracking Caribbean storms. He based these studies on the analysis of seventy logbooks in the possession of George Blunt, publisher of *The American Coast Pilot, etc.*; see "Remarks on the Prevailing Storms of the Atlantic Coast of the North American States; by William C. Redfield, of the City of New York," *American Journal of Science* 20 (1831): 17−51. For discussion of Redfield, see Gisela Kutzbach, *The Thermal Theory of Cyclones: A History of Meteorological Thought in the Nineteenth Century* (Boston: American Meteorological Society, 1979); William Goetzman, *New Lands, New Men: America and the Second Great Age of Discovery* (New York: Viking, 1986), chap. 8; and Mark S. Monmonier, *Air Apparent: How Meteorologists Learned to Map, Predict, and Dramatize Weather* (Chicago: University of Chicago Press, 1999), 31−42. For a usefully detailed summary of the development and fate of Maury's charts, see Pinsel, "The Wind and Current Chart Series." More work might be done specifically on earlier instances of the systematic collation of maritime observations out of logbooks; since such efforts generally had strategic or commercial value, they were seldom attended with public fanfare. It was perhaps here, in the highly public nature of Maury's undertaking, that his program was most distinctive.

21. One heavily used copy I have examined is penciled over with period annotations, including a giant red pointing hand, a fingerboard indicating the "New Route" and noting that it shaved a thousand miles off the old (Maury Charts, "A Series" drawer, Library of Congress).

22. It should be stated here that we very much need a close investigation of the truth of these claims, which mostly derive from Maury's own writings or writings derived from them. In the printed discussion that followed the presentation of Leighly's "M. F. Maury in His Time," Harold Burstyn alluded to work by the economic historian Douglass North suggesting that Maury's charts actually had little impact on shipping. Unfortunately, I can find nothing to this effect in North's published work, and the proposition would seem to be belied by the interest shown by maritime insurance companies at the time, as well as the reports of the clipper races (see Hearn's useful summary in chap. 8 of *Tracks in the Sea*). Moreover, Edward L. Towle's often overlooked but very revealing study ("Science, Commerce and the Navy on the Seafaring Frontier" [Ph.D. diss., University of Rochester, 1966]) actually does examine Maury manuscript correspondence and concludes the charts were very much in demand. Additional archival work on the logbooks of merchant and whaling vessels in the period, and even research in contemporary newspapers, should make it possible to confirm (or quite possibly revise) the standard account of the usefulness of the track charts. See, for instance, the resources cited in Lewis J. Darter, "Federal Archives Relating to Matthew Fontaine Maury," *American Neptune* 1 (January 1941): 149−58.

23. For Maury's pilot charts, Maury Charts, "C Series," Library of Congress.

24. Maury alludes to a contemporary analogue computational device for use at sea, Piddington's Horn Cards (Maury, *The Physical Geography of the Sea,* ed. Leighly, 380), which were sets of moving wheels that could be used for calculating bearings with respect to various winds, a kind of sailing slide rule. See Henry Piddington, *The Sailor's Horn-Book for the Law of Storms: Being a practical exposition of the theory of the law of storms, and its uses to mariners of all classes in all parts of the world, shewn by transparent cards* (New York: Wiley, 1848). My thanks to Katie Anderson for this reference. For a broad treatment of such devices, see the recent book on volvelles, Jessica Helfand, *Reinventing the Wheel* (Princeton, NJ: Princeton Architectural Press, 2002).

25. See Octavius Howe and Frederick Matthews, *American Clipper Ships, 1833–1858,* 2 vols. (New York: Dover, 1986).

26. Matthew Fontaine Maury, *The Physical Geography of the Sea* (New York: Harper & Brothers, 1855), 293. See also Williams, *Matthew Fontaine Maury,* 190.

27. Maury wrote: "In overhauling the log-books for data for this chart, I have followed vessels with the water thermometer to and fro across the seas" (*The Physical Geography of the Sea,* ed. Leighly, 358).

28. Williams, *Matthew Fontaine Maury,* 532.

29. Maury was working with the Philadelphia firm of E. C. and J. Biddle, and it was they who advised him (via his nephew) to undertake a quick book for the general reader, lest "some Yankee bookmaker steal his thunder and reap a fortune from it. (For this conversation, see Williams, *Matthew Fontaine Maury,* 258−59.) Such maneuverings were familiar fare in the Philadelphia book market: it is quite possible that the Biddles had in mind the chop-shop piracy of Mary Somerville's *Preliminary Dissertation to the "Mechanism of the Heavens"* by Cary & Lea in the 1830s. (For this incident, see Elizabeth Chambers Patterson, *Mary Somerville and the Cultivation of Science, 1815–1840* [Boston: Martinus Nijhoff, 1983], 88−89).

30. For two valuable recent studies in this area, see James A. Secord, *Victorian Sensation: The Extraordinary Publication, Reception, and Secret Authorship of "Vestiges of the Natural History of Creation"* (Chicago: University of Chicago Press, 2001); and Adrian Johns, *The Nature of the Book: Print and Knowledge in the Making* (Chicago: University of Chicago Press, 2000). For an introduction to a diverse set of approaches to these problems, see the essays in *Books and the Sciences in History,* ed. Marina Frasca-Spada and Nick Jardine (Cambridge: Cambridge University Press, 2000).

31. I am here thinking in particular of Bernard Lightman's work on Victorian science and popular culture, as well as James Secord's new research on science and the press. For an excellent introduction to these issues, see Bernard Lightman, ed., *Victorian Science in Context* (Chicago: University of Chicago Press, 1997), particularly his own essay in the volume, "'The Voices of Nature': Popularizing Victorian Science."

32. Maury cites the familiar turn of phrase, "look through nature up to nature's God" (*The Physical Geography of the Sea,* ed. Leighly, 352). For a general discussion of natural theology, see David C. Lindberg and Ronald Numbers, eds., *God and Nature: Historical Essays on the Encounter between Christianity and Science* (Berkeley: University of California Press, 1986). For the more immediate Anglo-American context in the first half of the nineteenth century, see chap. 1 and the conclusion of Susan Faye Cannon,

Science in Culture (New York: Dawson; Science History Publications, 1978). While mostly concerned with a later period, Jon H. Robert, *Darwinism and the Divine in America: Protestant Intellectuals and Organic Evolution, 1859–1900* (Madison: University of Wisconsin Press, 1988), offers invaluable details on the competing varieties of natural theology current in nineteenth-century American intellectual life.

33. On Maury's religious life, including a discussion of his long, slow decision to be confirmed and receive Communion, see Williams, *Matthew Fontaine Maury,* 105–6, 125–26, 161, 339–42, 449, etc.

34. Maury, *The Physical Geography of the Sea,* ed. Leighly, 153.

35. Maury was not alone or absolutely original in this program, but he seems to have been its most rigorous and vigorous exponent in the period, and his metrical preoccupations made his treatment unique. It would be interesting to have a more complete study of the role of natural theological framings of oceanic and meteorological dynamics in this period. William Scoresby would be a significant point of departure. See his "Observations on the Currents and Animalcules of the Greenland Sea," *Edinburgh Philosophical Journal* 4 (1821): 111–14; and the general discussion of Scoresby by Alison Winter in her " 'Compasses All Awry': The Iron Ship and the Ambiguities of Cultural Authority in Victorian Britain," *Victorian Studies* 3 (1994): 69–98. Other relevant primary texts would include Philip Henry Gosse, *The Ocean* (London: Society for Promoting Christian Knowledge, 1846); Samuel G. Goodrich, *A Glance at the Physical Sciences; or, The Wonders of Nature, in Earth, Air, and Sky; by the author of Peter Parley's Tales* (Boston: Bradbury, Soden & Co., 1844); and Rosina Maria Zornlin, *The World of Waters; Or, Recreations in Hydrology* (London: John W. Parker, 1843). This last contains a particularly noteworthy chapter on the oceans, which concludes with a poetic invocation of providence.

36. I have recently played out Maury's clockwork metaphor for the sea in "Mapping Time," *Daedalus* 132, no. 2 (2003): 5–19.

37. Maury, *The Physical Geography of the Sea,* ed. Leighly, 223. In Maury's account, the "coral insects" helped regulate currents by continuously removing dissolved minerals from seawater; those currents ensured that the food of the whales drifted to northern waters.

38. I am here referring to Burstyn's entry in the *Dictionary of Scientific Biography* (s.v.); Leighly's "M. F. Maury in His Time," and his introduction to Maury's *The Physical Geography of the Sea,* ed. Leighly; and Reingold, "Two Views of Maury." For an interestingly back-handed explanation of Maury's prominent reputation in Europe, see Burstyn's comment at the Monaco conference in 1966, to the effect that Maury represented exactly the kind of American science that European scientists wanted to see: empirical and fact gathering (by implication, a second-rate operation). See *Bulletin de l'Institut Océanographique,* Fondation Albert I, Prince de Monaco, special no. 2 (1968): 161. This argument fit into the emerging attention to the Basalla model for "colonial science." But this reading, interesting as it is, would now need to confront the arguments that measurement-intensive earth sciences represented "the new great thing in professional science in the first half of the nineteenth century." See Cannon, "Humboldtian Science," in *Science in Culture,* 105 ff.

39. Dick offers a balanced discussion of this larger issue of Maury's changing fate at the hands of scholars and reaches his own conclusions on the basis of an assessment of

Maury's institutional obligations and constraints (*Sky and Ocean Joined,* esp. 62, 108–9, 116–17).

40. Note that positioning Maury as an American Humboldtian is very much what William Goetzmann has at least begun in *New Lands, New Men: American and the Second Great Age of Discovery* (New York: Viking, 1986), although I think a good deal remains to be said on this point, for example, on the graphical techniques of the charts, the effort to link biogeography to the physical features of the earth, etc. To take just one example, see the remarkable passage in Maury, *The Physical Geography of the Sea,* ed. Leighly, 278, where Maury links the layering of the atmosphere to a geological column and suggests that the wind strata, too, are "engraved" with history. Note also Maury's work as a mine technician, which would further have bound him to Humboldt (see chap. 3 in this volume).

41. A useful overview of recent developments in this area can be found in David C. Lindberg and Ronald L. Numbers, eds., *When Science and Christianity Meet* (Chicago: University of Chicago Press, 2003).

42. See Alain Corbin, *Le Territoire du Vide* (Paris: Aubier, 1988), esp. chap. 1.

43. See, for example, Charles Wilkes, *Narrative of the United States Exploring Expedition,* 5 vols. (Philadelphia: Lea & Blanchard, 1845), where he inserted a plea that something be done for the "improving of the condition of sailors, elevating them in their circumstances," so that the "tars" would cease to be "considered, as they now frequently are, worthless reprobates, opposed to every thing that is sacred" (5:502). On the moral significance of precise measurement, see generally: M. Norton Wise, ed., *The Values of Precision* (Princeton, NJ: Princeton University Press, 1995). For a rich Victorian case, see Simon Schaffer, "Metrology, Metrication, and Victorian Values," in *Victorian Science in Context,* ed. Lightman.

44. Maury, *The Physical Geography of the Sea,* ed. Leighly, 283; Maury, *The Physical Geography of the Sea* (1855), 96.

45. Maury, *The Physical Geography of the Sea,* ed. Leighly, 8. The same letter also reads:

> I am happy to contribute my mite toward furnishing you with material to work out still farther toward perfection your great and glorious task, not only of pointing out the most speedy routes for ships to follow over the ocean, but also of teaching us sailors to look about us, and see by what wonderful manifestations of the wisdom and goodness of the great God we are continually surrounded. For myself, I am free to confess that for many years I commanded a ship, and, although never insensible to the beauties of nature upon the sea or land, I yet feel that, until I took up your work, I had been traversing the ocean blindfolded, I did not think; I did not know the amazing and beautiful combination of all the works of Him whom you so beautifully term "the Great First Thought."

There is much more evidence for this moral-reform reading of *The Physical Geography of the Sea* scattered throughout the text.

46. Maury, *The Physical Geography of the Sea* (1855), 47.

47. The analogy grows richer when we remind ourselves that likening the earth system to a central heating system had a very different feel in the mid-nineteenth cen-

tury: the observatory's closed water system was one of the very earliest such installations in the United States, and in 1855, the year this passage first appeared, the British-trained Joseph Nason designed the heating and ventilation system for the new House and Senate additions to the U.S. capitol—central heating was not only much on American minds, it was cutting edge technology as well. See Eugene S. Ferguson, "An Historical Sketch of Central Heating: 1800-1860," in *Early Building in America,* ed. Charles E. Peterson (Radnor, PA: Chilton, 1976), 165-85. Note that Maury may have gotten the idea of likening the tropics to the earth's "furnace" from the work of Heinrich Wilhelm Dove.

48. Series F is the final set of Maury's documents as they are listed, not as they were made.

49. Note that here, too, he was competing with Wilkes, who had collated currents and whaling grounds in a chart published in "Currents and Whaling," chap. 12 of vol. 5, of his *Narrative of the United States.*

50. Maury, *The Physical Geography of the Sea* (1855), 146, emphasis mine.

51. William Scoresby Jr., *An Account of the Arctic Regions,* 2 vols. (Edinburgh: Archibald Constable & Co., 1820), 1:8-12.

52. Maury made this declaration in a letter published in *Hunt's Merchants' Magazine* 24, no. 1 (1851): 773.

53. Maury wrote personally to Humboldt of this biogeographical discovery (Maury to Humboldt, 6 September 1849, Maury Papers, vol. 3, Manuscripts Collection, Library of Congress).

54. The story of the preservation of the *Acushnet* log data can be found in Wilson Heflin, *Herman Melville's Whaling Years,* ed. Mary K. Bercaw Edwards and Thomas Farel Heffernan (Nashville: Vanderbilt University Press, 2004).

CHAPTER 8

1. Katherine Emma Manthorne, *Tropical Renaissance: North American Artists Exploring Latin America, 1839-1879* (Washington, DC: Smithsonian Institution Press, 1989); Kevin J. Avery, *Church's Great Picture: The Heart of the Andes* (New York: Metropolitan Museum of Art, 1993).

2. "Gleanings from the Natural History of the Tropics," *Quarterly Review* 118 (1865): 167, 183.

3. G. Hartwig, *The Tropical World: A Popular Scientific Account of the Natural History of the Animal Kingdoms in the Equatorial Region* (London: Longman, Green, 1863); cf. Pierre Gourou, *The Tropical World: Its Social and Economic Conditions and Its Future Status,* trans. E. D. Laborde (1947; reprint, London: Longman, 1953).

4. For the significance of these voyages, see Alan Frost, "New Geographical Perspectives and the Emergence of the Romantic Imagination," in *Captain Cook and His Times,* ed. Robin Fisher and Hugh Johnston (London: Croom Helm, 1979).

5. David Arnold, "India's Place in the Tropical World, 1770-1930," *Journal of Imperial and Commonwealth History* 26 (1998): 1-21.

6. Bernard Smith, *European Vision and the South Pacific,* 2d ed. (New Haven, CT, and London: Yale University Press, 1985), 56-80; Mildred Archer, *Early Views of In-*

dia: The Picturesque Journeys of Thomas and William Daniell, 1786–1794 (London: Thames & Hudson, 1980); Giles Tillotson, *The Artificial Empire: The Indian Landscapes of William Hodges* (Richmond, Surrey: Curzon Press, 2000).

7. William Roxburgh, *Plants of the Coast of Coromandel,* 3 vols. (London: George Nicol, 1795–1819).

8. James Annesley, *Researches into the Causes, Nature and Treatment of the Most Prevalent Diseases of India, and of Warm Climates Generally,* 2 vols. (London: Longman, Rees, 1828).

9. J. Forbes Royle, *Illustrations of the Botany and Other Branches of the Natural History of the Himalayan Mountains, and of the Flora of Cashmere,* 2 vols. (London: W. H. Allen, 1839).

10. "Canals of Irrigation in the N.W. Provinces," *Calcutta Review* 12 (1849): 79, 109.

11. For Hooker, see Leonard Huxley, *Life and Letters of Sir Joseph Dalton Hooker,* 2 vols. (London: John Murray, 1919); Ray Desmond, *Sir Joseph Dalton Hooker: Traveller and Plant Collector* (Woodbridge, Suffolk: Antique Collectors' Club, 1999).

12. James Emerson Tennent, *Ceylon: An Account of the Island, Physical, Historical and Topographical,* 2 vols. 3d ed. (London: Longman, Green, 1859), 1:99, describing his arrival at Galle in Ceylon in November 1845, remarked, "No traveller fresh from Europe will ever part with the impression left by his first gaze upon tropical scenery."

13. William Hepworth Dixon, review of *Notes of a Tour in the Plains of India, the Himala* [sic], *and Borneo,* pt. 1, by J. D. Hooker, *Athenaeum,* 21 October 1848, 1049–50.

14. "Dr Hooker's Botanical Mission to India," *Hooker's London Journal of Botany* 7 (1848): 315–16. The pantropical nature of this description is enhanced by the fact that neither pineapple nor breadfruit was native to Sri Lanka.

15. Jacques-Henri Bernardin de Saint-Pierre, *Paul and Virginia* (1788), trans. John Donovan (London: Peter Owen, 1982). The basis for this passage was Hooker's letter to Frances Henslow, 5 January 1848, Hooker's Indian Letters, Royal Botanic Gardens Kew Archives, London (hereafter RBG Kew Archives).

16. Alexander von Humboldt, *Cosmos: A Sketch of a Physical Description of the Universe,* trans. E. C. Otté, 2 vols. (London: George Bell, 1900), 2:373; J. D. Hooker, *Himalayan Journals; or, Notes of a Naturalist in Bengal, the Sikkim and Nepal Himalayas, the Khasi Mountains &c* (1854; reprint, London: Ward, Lock, 1891), xii.

17. Hooker, *Himalayan Journals,* 422.

18. Hooker to Elizabeth Rigby, 29 July 1848, Private Letters to J. D. Hooker, vol. 17, RBG Kew Archives.

19. Darwin to Hooker, 10 March 1854, in *The Correspondence of Charles Darwin,* vol. 5, *1847–1850,* ed Frederick Burkhardt and Sydney Smith (Cambridge: Cambridge University Press, 1989), 182; Hooker to Darwin, ca. 25 March 1854, in *The Correspondence of Charles Darwin,* 5:185.

20. Hooker, *Himalayan Journals,* 2.

21. Ibid., 51.

22. Ibid., 9–10, 17, 21, 36.

23. Ibid., 5.

24. Ibid., 101.

25. Hooker's ideas on plant ecology and species variation in the Himalayas were most fully developed in J. D. Hooker, "On the Climate and Vegetation of the Temperate and Cold Regions of East Nepal and the Sikkim Himalaya Mountains," *Journal of the Horticultural Society of London* 7 (1852): 69−131.

26. Hooker, *Himalayan Journals,* 85.

27. Many of Hooker's original drawings of rhododendron species, made on the spot, are preserved in the Library of the Royal Botanic Garden at Kew. These remain an important source of reference and identification. The sketches were later worked up by a professional botanical plant illustrator (W. H. Fitch) and published in Joseph Dalton Hooker, *The Rhododendrons of Sikkim-Himalaya* (London: Reeve, Benham & Reeve, 1849), and, with more attention to scenic context, *Illustrations of Himalayan Plants* (London: Lovell Reeve, 1855).

28. Hooker, *Himalayan Journals,* 85−6.

29. Ibid., 108.

30. Ibid., 137.

31. Ibid., 529.

32. B. H. Hogdson, "On the Colonization of the Himalaya by Europeans," in *Essays on the Languages, Literature, and Religion of Nepal and Tibet* (1858; reprint, London: Trubner, 1874), pt. 2:83−90; Hooker, *Himalayan Journals,* 70, 74, 81−82.

33. Hooker, *Himalayan Journals,* 87−88.

34. Smith, *European Vision,* 1−7.

35. Hooker, *Himalayan Journals,* 10−11, and notes in his Indian journal for 1 and 2 February 1848, RBG Kew Archives.

36. Hooker, *Himalayan Journals,* 164, 185, 267.

37. Joseph Dalton Hooker, *Flora Antarctica: The Botany of the Antarctic Voyage of H.M. Discovery Ships "Erebus" and "Terror" in the Years 1839−43,* 2 vols. (London: Reeve Brothers, 1847), but the description of Kerguelen's Island (2:219) hints at the search for a more powerful descriptive style.

38. James Paradis, "Darwin and Landscape," in *Victorian Science and Victorian Values: Literary Perspectives,* ed. James Paradis and Thomas Postlewait (New Brunswick, NJ: Rutgers University Press, 1985).

39. Hooker, *Himalayan Journals,* 72.

40. Ibid., 131. At the time, Hooker compared this view to Victoria Land in the Antarctic, "where the coloring of clouds & Snowy peaks exceeds anything I have seen or am likely to in the Himalayah," but he omitted this comparison from the published *Journals* (Hooker, Indian journal, 9 November 1848, RBG Kew Archives).

41. Huxley, *Life and Letters,* 1:267.

42. In his private correspondence, Hooker was skeptical about the ability of any painter to communicate the scale and complexity of a scene like that presented by the Himalayas (Hooker to Elizabeth Rigby, 29 July 1848, Private Letters to J. D Hooker, vol. 7, RBG Kew Archives).

43. Edwin Lankester, review of *Himalayan Journals,* by J. D. Hooker, *Athenaeum* 25 February 1854, 238.

44. Hooker, *Himalayan Journals,* 186−87. J. M. W. Turner's painting *Rain, Steam*

and Speed—the Great Western Railway was exhibited at the Royal Academy in London in 1844 and *The Whalers* in 1845.

45. Darwin to Hooker, 1 March 1854, in *Correspondence of Charles Darwin,* ed Burkhardt and Smith, 5:179.

46. Review of *Himalayan Journals,* by J. D. Hooker, *Spectator* 4 March 1854, 253.

CHAPTER 9

1. Both photographs belong to a selection of 179 images by Davis and Tattersall assembled by Derek Freeman for the Alexander Turnbull Library, Wellington, in 1946, now cataloged as PA Coll-3062. Besides these, the Alexander Turnbull Library has a very large and diverse collection of Samoan photographs. Some of the prints of the crowd of people on the veranda at Vailima are cropped further to accentuate the closeness of photographer and his subjects.

2. Henry Adams to Elizabeth Cameron, 15 December 1890, in *Henry Adams: Selected Letters,* ed. Ernest Samuels (Cambridge, MA, and London: Harvard University Press, Belknap Press, 1992), 231.

3. Evelyn Wareham, "Developing Samoan Difference, 1900–1914," *Turnbull Library Record: Pacific Studies Issue* 32 (1999): 57–74.

4. For an account of Apia during the nineteenth century, see Caroline Ralston, *Grass Huts and Warehouses: Pacific Beach Communities of the Nineteenth Century* (Canberra: Australian National University Press, 1977). See, too, Damon Salesa, " 'Troublesome Half-Castes': Tales of a Samoan Borderland" (Ph.D. diss., University of Auckland, 1997).

5. See Salesa, " 'Troublesome Half-Castes.' "

6. Paul M. Kennedy, *The Samoan Tangle: A Study in Anglo-German-American Relations, 1878–1900* (Dublin: Irish University Press, 1974). For the politics of Samoa and the interrelationships with various European nations in the nineteenth century, see also R. P. Gilson, *Samoa, 1830–1900: The Politics of a Multi-Cultural Community* (Melbourne and New York: Oxford University Press, 1970); and Malama Meleisea, *The Making of Modern Samoa: Traditional Authority and Colonial Administration in the Modern History of Samoa* (Suva: Institute of Pacific Studies, University of the South Pacific, 1987).

7. Meleisea, *The Making of Modern Samoa,* 40.

8. For late nineteenth- to early twentieth-century photography of Samoa, see, for example, Alison Devine Nordstrom, "Images of Paradise: Photographs of Samoa, 1880–1930" (Master's thesis, University of Oklahoma, 1990), and "Early Photography in Samoa: Marketing Stereotypes of Paradise," *History of Photography* 15 (1991): 272–86; Casey Blanton, ed., *Picturing Paradise: Colonial Photography of Samoa, 1875–1925* (Daytona Beach, FL: Daytona Beach Community College, 1995); Peter Mesenhöller and Jutte-Beata Engelhard, eds., *Bilder aus dem Paradies: Koloniale Fotografie aus Samoa, 1875–1925* (Marburg: Jonas, 1995); Anne Maxwell, *Colonial Photography and Exhibitions: Representations of "Native" People and the Making of European Identities* (London and New York: Leicester University Press, 1999); Gordon Maitland, "The Two Sides of the Camera Lens: Nineteenth-Century Photography and the Indigenous People of the Southwest Pacific," *Photofile: South Pacific* 6, no. 3 (1988): 47–59; Judy Annear,

ed., *Portraits of Oceania* (Sydney: Art Gallery of New South Wales, 1997); Christaud Geary and Virginia-Lee Webb, eds., *Delivering Views: Distant Cultures in Early Postcards* (Washington, DC: Smithsonian Institution Press, 1998).

9. See, for example, Maxwell, *Colonial Photography and Exhibitions;* Salesa, " 'Troublesome Half-Castes.' "

10. See, for example, Leonard Bell, "To See or Not to See: Conflicting Eyes in the Travel Art of Augustus Earle," in *Orientalism Transposed: The Impact of the Colonies on British Culture,* ed. Julie F. Codell and Dianne Sachko Macleod (Aldershot and Brookfield, VT: Ashgate, 1998), "Looking at Goldie: Face to Face with *'All 'e Same T'e Pakeha,*" in *Double Vision: Art Histories and Colonial Histories in the Pacific,* ed. Nicholas Thomas and Diane Losche (Cambridge: Cambridge University Press, 1999), and "Colonial Eyes Transformed: Looking at/in Paintings, an Exploratory Essay," *Australian and New Zealand Journal of Art* 1, no. 1 (2000): 40 − 62; Deborah Poole, *Vision, Race and Modernity: A Visual Economy of the Andean Image World* (Princeton, NJ: Princeton University Press, 1997); Elizabeth Edwards, *Raw Histories: Photographs, Anthropology and Museums* (Oxford and New York: Berg, 2001); and Nancy Leys Stepan, *Picturing Tropical Nature* (London: Reaktion, 2001). Deborah Poole notes (219) that it can be problematic to assume a silent or passive indigenous subject even in the most repressive acts of photography.

11. See, for example, in respect to photographs of people in colonized places, Mieke Bal's comments on Malek Alloula's *The Colonial Harem* (Minneapolis: University of Minnesota Press, 1986) in her *Double Exposures: The Subject of Cultural Analysis* (New York and London: Routledge, 1996).

12. The description of Davis can be found in George Egerton Westbrook, "Samoan Personalities" (n.d.), MS Papers 5498-1, n.p., Alexander Turnbull Library, Wellington. This manuscript and Westbrook's "Reminiscences" ([n.d.], MS Papers 61, Alexander Turnbull Library, Wellington) provide the fullest account of Davis's personality, behavior, and activities besides photography in Samoa; on Davis and the Sydney Photographic Company, see Sandy Barrie, "Australians behind the Camera: Early Australian Photographers" (1992), 2, N-Z, 157 (typescript held in the Alexander Turnbull Library, Wellington); Alan Davies and Peter Stitchbury, *The Mechanical Eye in Australia: Photography, 1840 − 1900* (Melbourne: Oxford University Press, 1985), 112. These works list a "J. Davis," a photographer, at several addresses in Sydney from 1800 to 1873. Davis's origins and activities in Australia are still to be clarified. Dirk Spenneman (Charles Sturt University), who is putting together a Web exhibit on Davis's use of an unusual bisected stamp in 1895 during the course of his work as the postal services provider in Apia from 1886 to 1900, believes that Davis possibly worked in the mint in Sydney, too (e-mail communication, 2 May 2002). Thanks to Dirk Spenneman, also, for determining Davis's year of death (1903), which had hitherto generally been given as 1895.

13. For descriptions of Davis's and Tattersall's commercial studio activities and work in Apia, see Nordstrom, "Images of Paradise," and "Early Photography in Samoa."

14. The eight Cusack-Smith albums in the Alexander Turnbull Library show diplomatic, social, and commercial activities of the Samoan and colonial communities and include many photographs by Davis, Tattersall, Andrew, and Cusack-Smith himself.

15. Augustin Kramer, *The Samoan Islands,* trans. Theodore Verhaaren, 2 vols. (1901; reprint, Auckland: Polynesian Press, 1994), 1:xviii.

16. Melissa Banta and Curtis Hinsley, *From Site to Sight: Anthropology, Photography and the Power of Imagery* (Cambridge, MA: Peabody Museum Press, 1986); Elizabeth Edwards, ed., *Anthropology and Photography, 1860–1920* (New Haven, CT, and London: Yale University Press, 1992), esp. Edwards's introduction; Peter Mesenhöller, "Ethnography Considers History: Some Examples from Samoa," in *Picturing Paradise,* ed. Blanton. See also Edwards, *Raw Histories.*

17. Nordstrom, "Early Photography in Samoa," 278.

18. Maxwell, *Colonial Photography and Exhibitions,* 166.

19. A notable exception to this is Elizabeth Edwards, "Time and Space on the Quarter Deck: Two Samoan Photographs by Captain W. Acland," in *Raw Histories.*

20. Nordstrom, "Photography of Samoa: Production, Dissemination and Use," in *Picturing Paradise,* ed. Blanton, 27.

21. A photograph in the Charles Kay Ward album (PA1-q-736, Alexander Turnbull Library, Wellington) is one of those that shows spectators to some extent.

22. Conventionally, in most of the work of colonial and traveling photographers and painters, a strict separation between observer and the observed was maintained, with the maker of the representation kept "absent" and signs of photographers' or painters' presence kept out of the picture. See, for example, Bell, "To See or Not to See"; and Edwards, *Raw Histories.*

23. The social history of clowning in Samoa is noted by Andrew Ross, "Cultural Preservation in the Polynesia of the Latter Day Saints," in *The Chicago Gangster Theory of Life: Nature's Debt to Society* (London and New York: Verso, 1999), 62.

24. For a detailed account of Samoan comic sketches, see Caroline Sinavaiana, "Where the Spirits Laugh Last: Comic Theatre in Samoa," in *Clowning as Critical Practice: Performance Humour in the South Pacific,* ed. William E. Mitchell (Pittsburgh: University of Pittsburgh Press, 1992), 192–219. Elizabeth Edwards has observed, too, that it would be a mistake to see indigenous reenactments of apparently traditional or customary practices for the purposes of photography as passive—i.e., subject to the total control and direction of the colonial photographer (*Raw Histories,* 172).

25. Cusack-Smith albums, Alexander Turnbull Library, Wellington: albums 1–2 (PAI-o-543 to 545), album 3 (PAI-q-274), and albums 4–8 (PAI-o-545–49).

26. Manuel Alvarado, "Photographs and Narrativity," *Screen Education,* nos. 32–33 (Autumn–Winter 1979–80): 9, argued that the authority of the Western photographer could be undermined by showing the photograph of a non-Western subject being produced, thus negating the "conventional effacement of its [the photograph's] own production" and revealing the photograph to be an artifact rather than a bit of "reality" neutrally recorded.

27. George Egerton Westbrook, "Reminiscences," MS papers 61, folder 65: "I assist a Photographer," 4, Alexander Turnbull Library, Wellington.

28. The phrase "star picture" is from ibid.

29. Salesa, "'Troublesome Half-Castes,'" 177.

30. Note Georges Perec, "The Street," in *Species of Spaces and Other Pieces* (London: Penguin, 1997), 47: "Contrary to the buildings, which almost always belong to someone, the streets in principle belong to no-one."

31. Besides the two street scenes discussed in this chapter, the Alexander Turnbull Library collection includes several other such photographs from the Davis / Tattersall Studio—for example, one with a group of soldiers making up the American guard (PA Coll-3062-1-03), and another, with an ethnically mixed group, simply titled, *Street Scene, Apia* (PA Coll-3062-1-07).

32. *Scene at Matafele* (alternatively titled *Old View of Matafele*), Cusack-Smith album 5, Alexander Turnbull Library, Wellington. The precise date and authorship of this photograph are uncertain.

33. Cusack-Smith album 6, Alexander Turnbull Library, Wellington. Authorship of this photograph is unclear.

34. That the act of photography itself penetrates has been asserted or implied by some writers—for example, Susan Sontag, *On Photography* (New York: Farrar, Straus & Giroux, 1978); and Fredric Jameson, *Signatures of the Visible* (London and New York: Routledge, 1990). Jameson writes of representational "products" that "you can possess visually": "The visual is essentially pornographic, which is to say that it has as its end in rapt, mindless fascination" (1).

35. Peter Mesenhöller, "Ethnography Considers History: Some Examples from Samoa," in *Bilder aus dem Paradies,* ed. Mesenhöller and Engelhard, 46.

36. For examples of the standard or conventional photographs of *taupou,* see Davis, *Vao, taupou of Apia* (ca.1893), reproduced in *Bilder aus dem Paradies,* ed. Mesenhöller and Engelhard, 75; and Tattersall, *Taupou* (ca.1905) and Thomas Andrew, *Taupou,* reproduced in *Picturing Paradise,* ed. Blanton, 124 and 127, respectively.

37. Meleisea, *The Making of Modern Samoa,* 21.

38. Kramer, *The Samoan Islands,* 2:113, plate 39; and file note for the photograph, reference number PA1-0-547-47-1, Alexander Turnbull Library, Wellington.

39. The same photograph, as reproduced in Kramer's book, is cropped differently, so that the figure appears full-standing, not as close as the figure in the photograph in the Cusack-Smith album. The manner in which a photograph is cropped can, of course, crucially affect both how the photograph is seen and what it can do. That the figure in Kramer's book is a little more distant fits more the function intended for the photograph—as a component in a project of anthropological record.

40. On images of young indigenous women as standing for the desired land, see, for example, Nicholas Monti, *Africa Then: Photographs, 1840 – 1910* (London: Thames & Hudson, 1987), 56. Monti writes: "The relationship between the white man and the native woman was often perceived as that between the colonizer and the colonized country, in which exotic transgression is interpreted as a return to the immediacy of passion that the Western world had long repressed. . . . The seduction and conquest of the African woman became a metaphor for the conquest of Africa itself. A powerful erotic symbolism linked a woman's femininity so strongly to the attraction of the land that they became a single idea." See also Abigail Solomon-Godeau, "Going Native," *Art in America* 77 (1989): 124, writing about Gauguin and Tahiti. Rod Edmond has noted how Polynesian islands have regularly been marked as "feminine," there to be "taken" (*Representing the South Pacific: Colonial Discourse from Cook to Gauguin* [Cambridge: Cambridge University Press, 1997], 74). This image persists in touristic representations today.

41 See, for example, Banta and Hinsley, *From Site to Sight;* and Edwards, *Raw Histories.*

42. The quote is John Tagg's formulation in *The Burden of Representation: Essays on Photographies and Histories* (Basingstoke: Macmillan Education, 1988), 64. Tagg asserts that frontality was fundamental to projects of photographic surveillance. However, the dynamics of the returned look or "gaze" in photography, and visual representations generally, can be very complex and not reducible to a single level of operation. See Bell, "Face to Face with Goldie." See also Catherine Lutz and Jane Collins, "The Photograph as an Intersection of Gazes: The Example of the *National Geographic,*" in *Visualising Theory: Selected Essays from V.A.R., 1990−1994,* ed. Lucien Taylor (New York and London: Routledge, 1996), 367−84, who postulate a typology of seven gazes, including the "gaze returned," to be found in photographic representations.

43. Kerry Howe, *Nature, Culture and History: The "Knowing" of Oceania* (Honolulu: University of Hawai'i Press, 2000), 85.

44. See, for example, James R. Ryan, *Picturing Empire: Photography and the Visualisation of the British Empire* (London: Reaktion Books, 1997); and Maxwell, *Colonial Photography and Exhibitions.*

45. Salesa, "'Troublesome Half-Castes.'" For example, Salesa, drawing on Gloria Anzaldúa, *Borderlands: The New Mestiza = La Frontera* (San Francisco: Aunt Lute, 1987), 174, characterizes a "borderland" as not merely an "ill-defined, indeterminate place, but a space that defines those things that lie beyond it or on either side . . . a site where opposing sides contest their boundaries."

CHAPTER 10

1. William Wordsworth, *Poems,* ed. John O. Hayden (Harmondsworth: Penguin, 1977), 1:578.

2. *Oxford English Dictionary,* s.v. "tropicality," http://dictionary.oed.com.

3. Mark Harrison, "'The Tender Frame of Man': Disease, Climate, and Racial Difference in India and the West Indies, 1760−1860," *Bulletin of the History of Medicine* 70 (1996): 70−75.

4. Benjamin Moseley, *A Treatise on Tropical Diseases; on Military Operations; and on the Climate of the West-Indies,* 3d ed. (London: T. Cadell, 1792), 102.

5. Ibid., 113.

6. Ibid., 103.

7. James Lind, *An Essay on Diseases Incidental to Europeans in Hot Climates,* 3d ed. (London: T. Becket, 1777), 2.

8. Ibid., 15.

9. Philip D. Curtin, *Death by Migration: Europe's Encounter with the Tropical World in the Nineteenth Century* (Cambridge: Cambridge University Press, 1989).

10. James Johnson and James Ranald Martin, *The Influence of Tropical Climates on European Constitutions,* 6th ed., "revised and greatly improved" (London: S. Highley, 1841), 2. This title was originally published by James Johnson in 1813. Martin became its cowriter in 1841 and its author (after Johnson's death) in 1845.

11. John Tregenza, *Professor of Democracy: The Life of Charles Henry Pearson, 1830−*

1894, Oxford Don and Australian Radical (Melbourne: Melbourne University Press, 1968), 2-3.

12. Charles H. Pearson, *National Life and Character: A Forecast* (London: Macmillan, 1893), 16.

13. Tregenza, *Professor of Democracy,* 234.

14. Pearson, *National Life and Character,* 16.

15. Ibid., 64.

16. Ibid., 85.

17. Ibid., 14.

18. Tregenza, *Professor of Democracy,* 231.

19. Quoted in Marilyn Lake, "On Being a White Man, Australia, circa 1900," in *Cultural History in Australia,* ed. Hsu-Ming Teo and Richard White (Sydney: University of New South Wales Press, 2003), 104.

20. Ibid. See also Marilyn Lake, "White Man's Country: The Trans-National History of a National Project," *Australian Historical Studies* 122 (October 2003): 346-63, and "The White Man under Siege: New Histories of Race in the Nineteenth Century and the Advent of White Australia," *History Workshop Journal* 58 (Autumn 2004): 41-62.

21. Tregenza, *Professor of Democracy,* 232.

22. Sven Lindqvist, *"Exterminate All the Brutes"* (London: Granta, 2002), 78-79, 138-39.

23. Benjamin Kidd, *The Control of the Tropics* (New York and London: Macmillan, 1893), 3.

24. Ibid., 8-11, 14.

25. Ibid., 6.

26. Ibid., 54.

27. Ibid., 73.

28. Ibid., 78-80.

29. Charles E. Woodruff, *The Effects of Tropical Light on White Men* (New York and London: Rebman, 1905), 1.

30. Ibid., 3.

31. Rudyard Kipling, *Kim* (1901; reprint, Harmondsworth: Penguin, 1987), 173.

32. Woodruff, *The Effects of Tropical Light,* 4.

33. Jack London, "The Unparalleled Invasion," in *Jack London,* ed. Arthur Calder-Marshall (London: Bodly Head, 1963).

34. L. Marks and Michael Worboys, introduction to *Migrants, Minorities and Health: Historical and Contemporary Studies,* ed. L. Marks and Michael Worboys (London: Routledge, 1997), 9.

35. For example, David Arnold, ed., *Imperial Medicine and Indigenous Societies* (Oxford: Oxford University Press, 1989); Michael Worboys, "Colonial Medicine," in *Medicine in the Twentieth Century,* ed. Roger Cooter and John Pickstone (Amsterdam: Harwood, 2000).

36. For a much fuller discussion of this history, see Nancy Leys Stepan, *Picturing Tropical Nature* (London: Reaktion Books, 2001), chap. 5.

37. Patrick Manson, *Tropical Diseases: A Manual of the Diseases of Warm Climates* (London: Cassell & Co., 1898), xi.

38. Ibid., 387.

39. Ibid., 390, 412.

40. Ibid., 386.

41. Stepan, *Picturing Tropical Nature,* 176−78.

42. Ibid., 173.

43. Manson, *Tropical Diseases,* 416.

44. The illustration that Stepan discusses was not included until the 4th ed. of *Tropical Diseases* in 1907. She suggests it might be someone from the Middle East, but he looks European to me.

45. This photograph had been reproduced from A. Hansen and C. Looft, *Leprosy: In Its Clinical and Pathological Aspects,* trans. Norman Walker (Bristol: John Wright 1895), where it is described as illustrating tuberous leprosy of six years duration.

46. Michael Worboys, "Tropical Diseases," in *Companion Encyclopedia of the History of Medicine,* ed. W. F. Bynum and Roy Porter, 2 vols. (London and New York: Routledge, 1993), 1:512−13.

47. Laura Otis, *Membranes: Metaphors of Invasion in Nineteenth-Century Literature, Science, and Politics* (Baltimore and London: Johns Hopkins University Press, 2000).

48. *Report on Leprosy by the Royal College of Physicians, Prepared for Her Majesty's Secretary of State for the Colonies* (London: Eyre & Spottiswoode, 1867).

49. Morrell Mackenzie, "The Dreadful Revival of Leprosy," *Nineteenth Century* 23 (1889): 925−41.

50. I am grateful to Graham Burnett of Princeton University for the latter suggestion (personal communication, 12 July 2002). The quote is from William Shakespeare, *Richard II,* 2.1.

51. *British Medical Journal* 1 (1887): 1269−70; 2 (1887): 799−800; 1 (1888): 1398; 2 (1888): 1241.

52. *Times* (London), 13 June 1889, 4.

53. *Pall Mall Gazette,* 13 June 1889.

54. Otis, *Membranes,* 34.

55. *Times* (London), 19, 20, 21, and 22 June 1889.

56. *Pall Mall Gazette,* 20 June 1889.

57. *Times* (London), 22 and 24 June 1889; 1 and 5 July 1889.

58. *British Medical Journal* 1 (1898): 392.

59. Charles Dickens, *Hard Times* (Harmondsworth: Penguin, 1969), 65.

60. See Felix Driver, *Geography Militant: Cultures of Exploration and Empire* (Oxford: Blackwell, 2001), 171−80, for a detailed discussion of the relation between Booth's and Stanley's works.

61. William Booth, *In Darkest England and the Way Out* (London: Charles Knight, 1970), 14.

62. George Gissing, *The Nether World* (London: J. M. Dent, 1973), 8.

63. James Cantlie, *Degeneration amongst Londoners* (London: Field & Tuer, Leadenhall Press, 1885), 13−16.

64. Ibid., 20−23.

65. Ibid., 45.

66. Ibid., 52.

67. I am grateful to Mark Harrison for letting me see the proofs of his entry

on Cantlie for the *Oxford Dictionary of National Biography* (Oxford: Oxford University Press, 2004). On Cantlie and the London School of Tropical Medicine, see also G. C. Cook, *From the Greenwich Hulks to Old St. Pancras: A History of Tropical Disease in London* (London: Athlone Press, 1992).

68. James Cantlie, *Leprosy in Hong Kong* (Hong Kong: Kelly & Walsh, 1890), 1.

69. James Cantlie, *Report on the Conditions under Which Leprosy Occurs in China, Indo-China, Malaya, the Archipelago and Oceania* (London: Macmillan, 1897), 9.

70. Ibid., 125.

71. James Cantlie, *Physical Efficiency: A Review of the Deleterious Effects of Town Life upon the Population of Britain, with Suggestions for Their Arrest* (London and New York: G. P. Putnam's Sons, 1906), 19.

72. Cantlie, *Degeneration*, 8–13, 24–25. See also Mark Harrison's entry on Cantlie in the *Oxford Dictionary of National Biography.*

73. Cantlie, *Physical Efficiency,* xix–xx.

74. Jack London, "The People of the Abyss," in *Jack London: Novels and Social Writings* (New York: Library of America, 1982), 28.

75. Ibid., 22.

76. Ibid., 163–64.

77. Ibid., 181–82.

78. The phrase "disease germs" is from ibid., 30–31.

79. Oscar Wilde, *The Picture of Dorian Gray* (Oxford: Oxford University Press, 1982), 139.

80. Ibid., 224.

81. Ibid., 179.

82. Arthur Conan Doyle, *A Study in Scarlet* (Harmondsworth: Penguin, 2001), 7–8.

83. Arthur Conan Doyle, *The Complete Sherlock Holmes* (Harmondsworth: Penguin, 1981), 1012.

84. For a more detailed analysis of this story, see Rod Edmond, "'Without the Camp': Leprosy and Nineteenth-Century Writing," *Victorian Literature and Culture* 29 (2001): 2.

85. Driver, *Geography Militant;* Gareth Stedman Jones, *Outcast London: A Study in the Relationship between Classes in Victorian Society* (Oxford: Clarendon Press, 1971); Andrew Lees, *Cities Perceived: Urban Society in European and American Thought, 1820–1940* (Manchester: Manchester University Press, 1985); Daniel Pick, *Faces of Degeneration: A European Disorder, c. 1848–c. 1918* (Cambridge: Cambridge University Press, 1989).

86. Jones, *Outcast London,* 308–12.

87. Pick, *Faces of Degeneration,* 37–39.

88. Lees, *Cities Perceived,* 190–218, esp. 199.

89. D. H. Lawrence, "Nottingham and the Mining Countryside," in *Phoenix* (New York: Viking Press, 1974), 139: "The English character has failed to develop the real *urban* side of a man, the civic side. . . . The new cities of America are much more genuine cities, in the Roman sense, than is London or Manchester." Also, Raymond Williams, *The Country and the City* (London: Chatto & Windus, 1973).

90. Everard Digby, "The Extinction of the Londoner," *Contemporary Review* 86 (1904):123–24, cited in Lees, *Cities Perceived,* 199–200.

91. Luigi Sambon, "Remarks on the Possibility of the Acclimatisation of Europeans in Tropical Regions," *British Medical Journal* 1 (1897): 61−66.

92. *British Medical Journal* 1 (1897): 93.

93. *British Medical Journal* 1 (1897): 94.

CHAPTER 11

1. O. H. K. Spate, "'South Sea' to 'Pacific Ocean': A Note on Nomenclature," *Journal of Pacific History* 12, no. 4 (1977): 205−11.

2. J. Lennart Berggren and Alexander Jones, *Ptolemy's Geography: An Annotated Translation of the Theoretical Chapters* (Princeton, NJ, and Oxford: Princeton University Press, 2000), 6−7.

3. Ibid., 64.

4. Ibid., 67.

5. Ibid., 69.

6. Ibid., 70.

7. Arnold Guyot, *Physical Geography* (New York and Chicago: Ivison, Blakeman, Taylor & Co., 1885).

8. Ibid., 7.

9. Ibid., 71−73.

10. Ibid., 106.

11. Ibid., 114.

12. Ibid., 116.

13. Ibid., 118.

14. Ibid., 121.

15. Jean-François Staszak, *Géographies de Gauguin* (Paris: Editions Bréal, 2003).

16. Ibid., 85.

17. Claude Lévi-Strauss, *Tristes tropiques,* trans. John Weightman and Doreen Weightman (New York: Atheneum, 1974), 74.

18. Ibid., 78.

19. Ibid., 90.

20. Ibid., 330, 321.

21. Ibid., 333−34.

22. Miwon Kwon, "One Place after Another: Notes on Site Specificity," in *Space, Site, Intervention: Situating Installation Art,* ed. Erika Suderburg (Minneapolis and London: University of Minnesota Press, 2000).

Select Bibliography

Note. This bibliography is a select list of secondary works, mainly but not exclusively in English, concerned with the visualisation of the tropical world in science, cartography, medicine, literature, material culture, and the visual arts. It is not intended to be comprehensive. The list includes most of the works cited in this book but excludes primary sources referred to in the footnotes.

Agnew, Vanessa. 2003. "Pacific Island Encounters and the German Invention of Race." In *Islands in History and Representation,* edited by Rod Edmond and Vanessa Smith. London: Routledge.

Anderson, Katherine. 2005. *Predicting the Weather: Victorians and the Science of Meteorology.* Chicago: University of Chicago Press.

Anderson, Warwick. 1996. "Immunities of Empire: Race, Disease and the New Tropical Medicine." *Bulletin of the History of Medicine* 70:94–118.

———. 2003. *The Cultivation of Whiteness: Science, Health and Racial Destiny in Australia.* New York: Basic Books.

Annear, Judy, ed. 1997. *Portraits of Oceania.* Sydney: Art Gallery of New South Wales.

Anzaldúa, Gloria. 1987. *Borderlands: The New Mestiza = La Frontera.* San Francisco: Aunt Lute.

Aparicio, F. R., and S. Chávez-Silverman, eds. 1997. *Tropicalizations: Transcultural Representations of Latinidad.* Hanover, NH: University Press of New England.

Aravamudan, Srinivas. 1999. *Tropicopolitans: Colonialism and Agency, 1688–1804.* Durham, NC, and London: Duke University Press.

Archer, Mildred. 1980. *Early Views of India: The Picturesque Journey of Thomas and William Daniell, 1786–1794.* London: Thames & Hudson.

Arnold, David, ed. 1989. *Imperial Medicine and Indigenous Societies.* Oxford: Oxford University Press.

———. 1996. "Inventing Tropicality." In *The Problem of Nature: Environment, Culture and European Expansion.* Oxford: Blackwell.

———. 1996. *Warm Climates and Western Medicine: The Emergence of Tropical Medicine, 1600–1900.* Amsterdam: Rodopi.

———. 1998. "India's Place in the Tropical World, 1770–1930." *Journal of Imperial and Commonwealth History* 26:1–21.

Bann, Stephen. 1994. "Travelling to Collect: The Booty of John Bargrave and Charles Waterton." In *Travellers' Tales: Narratives of Home and Displacement,* edited by George Robertson, Melinda Mash, Lisa Tickner, Jon Bird, Barry Curtis, and Tim Putman. London: Routledge.

Banta, Melissa, and Curtis Hinsley. 1986. *From Site to Sight: Anthropology, Photography and the Power of Imagery.* Cambridge, MA: Peabody Museum Press.

Beer, Gillian. 1996. *Open Fields: Science in Cultural Encounter.* Oxford: Clarendon Press.

Bell, Leonard. 1998. "To See or Not to See: Conflicting Eyes in the Travel Art of Augustus Earle." In *Orientalism Transposed: The Impact of the Colonies on British Culture,* edited by Julie F. Codell and Dianne Sachko Macleod. Aldershot and Brookfield, VT: Ashgate.

———. 1999. "Looking at Goldie: Face to Face with *'All 'e Same T'e Pakeha'.*" In *Double Vision: Art Histories and Colonial Histories in the Pacific,* edited by Nicholas Thomas and Diane Losche. Cambridge: Cambridge University Press.

———. 2000. "Colonial Eyes Transformed: Looking at/in Paintings, an Exploratory Essay." *Australian and New Zealand Journal of Art* 1, no. 1:40–62.

Belluzzo, Ana Maria de Moraes, ed. 1995. *The Voyagers' Brazil.* 3 vols. São Paulo: Metalivros.

Birtles, T. G. 1997. "First Contact: Colonial European Preconceptions of Tropical Queensland Rainforest and Its People." *Journal of Historical Geography* 23:393–417.

Blanton, Casey, ed. 1995. *Picturing Paradise: Colonial Photography of Samoa, 1875–1925.* Daytona Beach, FL: Daytona Beach Community College.

Botting, Douglas. 1973. *Humboldt and the Cosmos.* London: Michael Joseph.

Bourguet, Marie-Noëlle, Christian Licoppe, and H. Otto Sibum, eds. 2002. *Instruments, Travel and Science: Itineraries of Precision from the Seventeenth to the Twentieth Century.* London and New York: Routledge.

Bowd, Gavin, and Daniel Clayton. 2003. "Fieldwork and Tropicality in French Indochina: Reflections on Pierre Gourou's *Les paysans du delta tonkinois,* 1936." *Singapore Journal of Tropical Geography* 24, no. 2:147–68.

Bowler, Peter. 2000. *The Earth Encompassed. A History of the Environmental Sciences.* New York and London: W. W. Norton & Co.

Bravo, Michael. 1996. "Ethnological Encounters." In *Cultures of Natural History,* edited by Nicholas Jardine, James A. Secord, and Emma C. Spary. Cambridge: Cambridge University Press.

———. 1999. "Precision and Curiosity in Scientific Travel: James Rennell and the Orientalist Geography of the New Imperial Age (1760–1830)." In *Voyages and Visions: Towards a Cultural History of Travel,* edited by J. Elsner and J.-P. Rubiés. London: Reaktion Books.

Browne, Janet. 1996. "Botany in the Boudoir and Garden: The Banksian Context." In *Visions of Empire: Voyages, Botany and Representations of Nature,* edited by D. P. Miller and P. H. Reill. Cambridge: Cambridge University Press.

Bruneau, M., and D. Dory, eds. 1989. *Les enjeux de la tropicalité.* Paris: Masson.

Burnett, D. G. 2000. *Masters of All They Surveyed: Exploration, Geography and a British El Dorado.* Chicago: University of Chicago Press.

———. 2003. "Mapping Time." *Daedalus* 132, no. 2:5–19.

Camerini, Jane. 1993. "Evolution, Biogeography, and Maps: An Early History of Wallace's Line." *Isis.* 84:700–27.

———. 1996. "Wallace in the Field." In *Science in the Field,* edited by Henrika Kuklick and Robert E. Kohler. Chicago: University of Chicago Press.

Cannon, S. F. 1978. *Science in Culture: The Early Victorian Period.* New York: Science History Publications.

Carney, Judith. 2001. *Black Rice: The African Origins of Rice Cultivation in the Americas.* Cambridge, MA, and London: Harvard University Press.

Carter, Paul. 1999. "Dark with Excess of Bright: Mapping the Coastlines of Knowledge." In *Mappings,* edited by Denis Cosgrove. London: Reaktion Books.

Cook, G. C. 1992. *From the Greenwich Hulks to Old St. Pancras: A History of Tropical Disease in London.* London: Athlone Press.

Cook, Ian, Philip Crang, and Mark Thorpe. 2004. "Tropics of Consumption." In *Geographies of Commodity Chains,* edited by Alex Hughes and Suzanne Reimer. London and New York: Routledge.

Cosgrove, Denis, ed. 1999. *Mappings.* London: Reaktion Books.

———. 2001. *Apollo's Eye: A Cartographic Genealogy of the Earth in the Western Imagination.* Baltimore: Johns Hopkins University Press.

Curtin, Philip D. 1964. *The Image of Africa: British Ideas and Action, 1780–1850.* Madison: University of Wisconsin Press.

———. 1985. "Medical Knowledge and Urban Planning in Tropical Africa." *American Historical Review* 90:594–613.

———. 1989. *Death by Migration: Europe's Encounter with the Tropical World in the Nineteenth Century.* Cambridge: Cambridge University Press.

Denevan, William. 2001. *Cultivated Landscapes of Native Amazonia and the Andes.* Oxford: Oxford University Press.

Dening, Greg. 1992. *Mr. Bligh's Bad Language: Passion, Power and Theatre on the Bounty.* Cambridge: Cambridge University Press.

Desmond, Ray. 1999. *Sir Joseph Dalton Hooker: Traveller and Plant Collector.* Woodbridge, Suffolk: Antique Collectors' Club.

Despoix, Philippe. 1996. "Naming and Exchange in the Exploration of the Pacific: On European Representations of Polynesian Culture in Late XVIII Century." In *Multiculturalism and Representation: Selected Essays,* edited by John Rieder and Larry E. Smith. Honolulu: University of Hawai'i Press.

Dettelbach, Michael. 1996. "Global Physics and Aesthetic Empire: Humboldt's Physical Portrait of the Tropics." In *Visions of Empire: Voyages, Botany and Representations of Nature,* edited by D. P. Miller and P. H. Reill. Cambridge: Cambridge University Press.

———. 1997. Introduction to *Cosmos: A Sketch of the Physical Description of the Uni-*

verse, by Alexander von Humboldt. Vol. 2. Baltimore, MD: Johns Hopkins University Press.

————. 1999. "The Face of Nature: Precise Measurement, Mapping and Sensibility in the Work of Alexander von Humboldt." *Studies in History and Philosophy of Biological and Biomedical Sciences* 30, no. 4:473–504.

Diener, Pablo, and Maria de Fátima da Costa. 2002. *Rugendas e o Brasil.* São Paulo: Capivara.

Douglas, Bronwen. 1999. "Art as Ethno-Historical Text: Science, Representation and Indigenous Presence in Eighteenth- and Nineteenth-Century Oceanic Voyage Literature." In *Double Vision: Art Histories and Colonial Histories in the Pacific,* edited by Nicholas Thomas and Diane Losche. Cambridge: Cambridge University Press.

————. 1999. "Science and the Art of Representing 'Savages': Reading 'Race' in Text and Image in South Seas Voyage Literature." *History and Anthropology* 11, nos. 2–3:157–201.

Douglas, Starr. 2004. "Natural History, Improvement and Colonisation: Henry Smeathman and Sierra Leone in the Late Eighteenth Century." PhD diss., Royal Holloway, University of London.

Drayton, Richard. 2000. *Nature's Government: Science, Imperial Britain, and the "Improvement" of the World.* New Haven, CT: Yale University Press.

Driver, Felix. 2001. *Geography Militant: Cultures of Exploration and Empire.* Oxford: Blackwell.

————. 2004. "Distance and Disturbance: Travel, Exploration and Knowledge in the Nineteenth Century." *Transactions of the Royal Historical Society,* 6th ser., 14:73–92.

————. 2004. "Imagining the Tropics: Views and Visions of the Tropical World." *Singapore Journal of Tropical Geography* 25, no. 1:1–17.

Driver, Felix, and Brenda Yeoh, eds. 2000. "Constructing the Tropics." *Singapore Journal of Tropical Geography* 21:1–98.

Driver, Felix, and Luciana Martins. 2002. "John Septimus Roe and the Art of Navigation." *History Workshop Journal* 54:144–61.

Duncan, James. 2000. "The Struggle to Be Temperate: Climate and 'Moral Masculinity' in Mid-Nineteenth Century Ceylon." *Singapore Journal of Tropical Geography* 21:34–47.

Dunn, C. 2002. "Tropicália, Counterculture and the Diasporic Imagination in Brazil." In *Brazilian Popular Music and Globalization,* edited by C. A. Perrone and C. Dunn. New York: Routledge.

Edmond, Rod. 1997. *Representing the South Pacific: Colonial Discourse from Cook to Gauguin.* Cambridge: Cambridge University Press.

————. 2001. "'Without the Camp': Leprosy and Nineteenth-Century Writing." *Victorian Literature and Culture* 29:507–18.

Edney, Matthew. 1997. *Mapping an Empire: The Geographical Construction of British India, 1765–1843.* Chicago: University of Chicago Press.

Edwards, Elizabeth, ed. 1992. *Anthropology and Photography, 1860–1920.* New Haven, CT, and London: Yale University Press.

————. 2001. *Raw Histories: Photographs, Anthropology and Museums.* Oxford and New York: Berg.

Eisenman, Stephen. 1997. *Gauguin's Skirt.* London: Thames & Hudson.

Ette, Ottmar. 2001. "'Eine Gemütsverfassung von moralischer Unruhe': Alexander von Humboldt und das Schreiben in der Moderne." In *Aufbruch in die Moderne*. Berlin: Akademie Verlag, 33–55.

Fabian, Johannes. 2000. *Out of Our Minds: Reason and Madness in the Exploration of Central Africa*. Berkeley: University of California Press.

Fauser, Hildegard. 1995. "O Barão Georg Heinrich von Langsdorff." In *O Brasil de Hoje no Espelho do Século XIX*, edited by Maria de Fátima G. Costa, Pablo Diener, and Dieter Strauss. São Paulo: Estação Liberdade.

Frost, Alan. 1979. "New Geographical Perspectives and the Emergence of the Romantic Imagination." In *Captain Cook and His Times*, edited by Robin Fisher and Hugh Johnston. London: Croom Helm.

Fussell, Paul. 1980. *Abroad: British Literary Traveling between the Wars*. Oxford: Oxford University Press.

Geary, Christaud, and Virginia-Lee Webb, eds. 1998. *Delivering Views: Distant Cultures in Early Postcards*. Washington, DC: Smithsonian Institution Press.

Gerbi, Antonello. [1955] 1973. *The Dispute of the New World: The History of a Polemic, 1750–1900*. Translated by Jeremy Moyle. Pittsburgh: University of Pittsburgh Press.

Gilson, R. P. 1970. *Samoa, 1830–1900: The Politics of a Multi-Cultural Community*. Melbourne and New York: Oxford University Press.

Glacken, Clarence J. 1967. *Traces on the Rhodian Shore: Nature and Culture in Western Thought from Ancient Times to the End of the Eighteenth Century*. Berkeley: University of California Press.

Godlewska, Anne. 1999. "From Enlightenment Vision to Modern Science? Humboldt's Visual Thinking." In *Geography and Enlightenment*, edited by David N. Livingstone and Charles W. J. Withers. Chicago: University of Chicago Press.

Gourou, Pierre. [1947] 1953. *The Tropical World: Its Social and Economic Conditions and Its Future Status*. Translated by E. D. Laborde. London: Longman.

Greenblatt, Stephen. 1991. *Marvellous Possessions: The Wonder of the New World*. Oxford: Oxford University Press.

Grove, Richard. 1992. "Origins of Western Environmentalism." *Scientific American*, July, 42–47.

———. 1995. *Green Imperialism: Colonial Expansion, Tropical Island Edens and the Origins of Environmentalism, 1600–1860*. Cambridge: Cambridge University Press.

Guelke, J. Kay, and Karen Morin. 2001. "Gender, Nature, Empire: Women Naturalists in Nineteenth-Century British Travel Writing." *Transactions of the Institute of British Geographers* 26:306–26.

Guest, Harriet. 2003. "Cook in Tonga: Terms of Trade." In *Islands in History and Representation*, edited by Rod Edmond and Vanessa Smith. London and New York: Routledge.

Hair, P. E. H., Adam Jones, and Robin Law, eds. 1992. *Barbot on Guinea. The Writings of Jean Barbot on West Africa, 1678–1712*. London: Hakluyt Society.

Harrison, Mark. 1992. "Tropical Medicine in Nineteenth-Century British India." *British Journal for the History of Science*. 25:299–318.

———. 1996. "'The Tender Frame of Man': Disease, Climate, and Racial Difference in India and the West Indies, 1760–1860." *Bulletin of the History of Medicine* 70:68–93.

———. 1999. *Climates and Constitutions: Health, Race, Environment and British Imperialism in India, 1600–1850.* New Delhi: Oxford University Press.

Hartley, Beryl. 1996. "The Living Academies of Nature: Scientific Experiment in Learning and Communicating the New Skills of Early Nineteenth-Century Landscape Painting." *Studies in the History and Philosophy of Science.* 27, no. 2:149–80.

Haynes. Douglas. 2001. *Imperial Medicine: Patrick Manson and the Conquest of Tropical Disease.* Philadelphia: University of Pennsylvania Press.

Hearn, Chester G. 2002. *Tracks in the Sea: Matthew Fontaine Maury and the Mapping of the Oceans.* Camden, ME: International Marine.

Holl, Frank, and Kai Reschke, eds. 1999. *Alexander von Humboldt: Netzwerke des Wissens.* Berlin: Haus der Kulturen der Welt.

Howe, Kerry. 2000. *Nature, Culture and History: The "Knowing" of Oceania.* Honolulu: University of Hawai'i Press.

Howse, Derek. 1979. *Greenwich Time and the Discovery of the Longitude.* Oxford: Oxford University Press.

Hulme, Peter. 1986. *Colonial Encounters: Europe and the Native Caribbean, 1492–1797.* London: Methuen.

———. 1990. "The Spontaneous Hand of Nature: Savagery, Colonialism, and the Enlightenment." In *The Enlightenment and Its Shadows,* edited by Peter Hulme and Ludmilla Jordanova. London: Routledge.

———. 1993. "Making Sense of the Native Caribbean." *New West Indian Guide* 67, nos. 3–4:189–220.

———. 2003. "Black, Yellow, and White on St. Vincent: Moreau de Jonnès's Carib Ethnography." In *The Global Eighteenth Century,* edited by Felicity A. Nussbaum. Baltimore: Johns Hopkins University Press.

———. 2005. "*The Fatal Gift:* storie di fine impero." In *Le maschere letterare l'impero britannico,* edited by Daniela Corona and Elio di Piazza. Pisa: Edizioni ETS.

Jacobs, Michael. 1995. *The Painted Voyage: Art, Travel and Exploration, 1564–1875.* London: British Museum.

Joppien, Rüdiger, and Bernard Smith, eds. 1985–88. *The Art of Captain Cook's Voyages.* 3 vols. New Haven, CT, and London: Yale University Press.

Kemp, Martin. 1990. *The Science of Art: Optical Themes in Western Art from Brunelleschi to Seurat.* New Haven, CT, and London: Yale University Press.

Kennedy, Dane. 1990. "The Perils of the Midday Sun: Climatic Anxieties in the Colonial Tropics." In *Imperialism and the Natural World,* edited by John M. MacKenzie. Manchester: Manchester University Press.

Kennedy, Paul M. 1974. *The Samoan Tangle: A Study in Anglo-German-American Relations, 1878–1900.* Dublin: Irish University Press.

Kenny, Judith T. 1995. "Climate, Race and Imperial Authority: The Symbolic Landscape of the British Hill Station in India." *Annals of the Association of American Geographers* 85:694–714.

Keynes, Richard, ed. 1979. *The Beagle Record.* Cambridge: Cambridge University Press.

King, Ronald, ed. 1981. *The Temple of Flora by Robert Thornton.* London: Weidenfeld & Nicolson.

Klein, Bernhard, and Gesa Mackenthun, eds. 2004. *Sea Changes: Historicizing the Ocean.* New York and London: Routledge.

Klonk, Charlotte. 1996. *Science and the Perception of Nature: British Landscape Art in the Late Eighteenth and Early Nineteenth Centuries*. New Haven, CT, and London: Yale University Press.

Kriz, Kay Dian. 2000. "Curiosities, Commodities, and Transplanted Bodies in Hans Sloane's 'Natural History of Jamaica.'" *William and Mary Quarterly* 57:35–78.

Kuklick, Henrika, and Robert E. Kohler, eds. 1996. *Science in the Field*. Chicago: University of Chicago Press.

Kutzbach, Gisela. 1979. *The Thermal Theory of Cyclones: A History of Meteorological Thought in the Nineteenth Century*. Boston: American Meteorological Society.

Lake, Marilyn. 2003. "White Man's Country: The Trans-National History of a National Project." *Australian Historical Studies* 122 (October): 346–63.

———. 2004. "The White Man under Siege: New Histories of Race in the Nineteenth Century and the Advent of White Australia." *History Workshop Journal* 58:41–62.

Lamb, Jonathan. 2001. *Preserving the Self in the South Seas, 1680–1840*. Chicago: University of Chicago Press.

Lange, Fritz, and Ilse Jahn, eds. 1973. *Die Jugendbriefe Alexander von Humboldts*. Berlin: Akademie Verlag.

Latour, Bruno. 1990. "Drawing Things Together." In *Representation in Scientific Practice*, edited by Michael Lynch and Steve Woolgar. Cambridge, MA: MIT Press.

———. 1999. "Circulating Reference: Sampling the Soil in the Amazon Forest." In *Pandora's Hope: Essays on the Reality of Science Studies*. Cambridge, MA: Harvard University Press.

Leask, Nigel. 2002. "Alexander von Humboldt and the Romantic Imagination of America: The Impossibility of Personal Narrative." In *Curiosity and the Aesthetics of Travel Writing, 1770–1840*. Oxford: Oxford University Press.

Lemmon, Kenneth. 1968. *The Golden Age of Plant Hunters*. London: Phoenix House.

Lepofsky, Dana. 1999. "Gardens of Eden? An Ethnocentric Reconstruction of Maohi (Tahitian) Cultivation." *Ethnohistory* 46, no. 1:1–29.

Lévi-Strauss, Claude. 1974. *Tristes tropiques*. Translated by John Weightman and Doreen Weightman. New York: Atheneum.

Lightman, Bernard, ed. 1997. *Victorian Science in Context*. Chicago: University of Chicago Press.

Lindqvist, Sven. 2002. *"Exterminate All the Brutes."* London: Granta.

Livingstone, David N. 1991. "The Moral Discourse of Climate: Historical Considerations of Race, Place and Virtue." *Journal of Historical Geography* 17:413–34.

———. 1992. *The Geographical Tradition: Episodes in the History of a Contested Enterprise*. Oxford: Blackwell.

———. 1999. "Tropical Climate and Moral Hygiene: The Anatomy of a Victorian Debate." *British Journal for the History of Science* 32:93–100.

———. 2002. "Tropical Hermeneutics and the Climatic Imagination." In *Science, Space and Hermeneutics: Hettner-Lecture, 2001*. Heidelberg: Department of Geography, University of Heidelberg.

Livingstone, David N., and Charles W. J. Withers, eds. 1999. *Geography and Enlightenment*. Chicago: University of Chicago Press.

Lutz, Catherine, and Jane Collins. 1996. "The Photograph as an Intersection of

Gazes: the Example of the *National Geographic.*" In *Visualising Theory: Selected Essays from V.A.R., 1990–1994,* edited by Lucien Taylor. New York and London: Routledge.

McEwan, Cheryl. 2000. *Gender, Geography and Empire: Victorian Women Travellers in West Africa.* Aldershot: Ashgate Publishing.

Mackay, David. 1985. *In the Wake of Cook: Exploration, Science and Empire, 1780–1801.* London: Croom Helm.

———. 1996. "Agents of Empire: The Banksian Collectors and Evaluation of New Lands." In *Visions of Empire: Voyages, Botany, and Representations of Nature,* edited by D. P. Miller and P. H. Reill. Cambridge: Cambridge University Press.

MacLeod, R., and M. Lewis, eds. 1988. *Disease, Medicine and Empire: Perspectives on Western Medicine and the Experience of European Expansion.* London: Routledge.

Maitland, Gordon. 1988. "The Two Sides of the Camera Lens: Nineteenth-Century Photography and the Indigenous People of the Southwest Pacific." *Photofile: South Pacific* 6, no. 3:47–59.

Manthorne, Katherine Emma. 1989. *Tropical Renaissance: North American Artists Exploring Latin America, 1839–1879.* Washington, DC: Smithsonian Institution Press.

Marks, Lara, and Michael Worboys. 1997. Introduction to *Migrants, Minorities and Health: Historical and Contemporary Studies,* edited by Lara Marks and Michael Worboys. London: Routledge.

Martins, Luciana. 1999. "Mapping Tropical Waters." In *Mappings,* edited by Denis Cosgrove. London: Reaktion Books.

———. 2000. "A Naturalist's Vision of the Tropics: Charles Darwin and the Brazilian Landscape." *Singapore Journal of Tropical Geography* 21, no. 1:19–33.

———. 2001. *O Rio de Janeiro dos Viajantes: O Olhar Britânico, 1800–1850.* Rio de Janeiro: Jorge Zahar.

———. 2004. "The Art of Tropical Travel, 1768–1830." In *Georgian Geographies,* edited by Miles Ogborn and Charles W. J. Withers. Manchester: Manchester University Press.

Mawer, Granville Allen. 1999. *Ahab's Trade: The Saga of South Sea Whaling.* New York: St. Martins.

McMinn, W. G. 1970. *Allan Cunningham: Botanist and Explorer.* Victoria: Melbourne University Press.

Meleisea, Malama. 1987. *The Making of Modern Samoa: Traditional Authority and Colonial Administration in the Modern History of Samoa.* Suva: Institute of Pacific Studies, University of the South Pacific.

Mesenhöller, Peter, and Jutte-Beata Engelhard, eds. 1995. *Bilder aus dem Paradies: Koloniale Fotografie aus Samoa, 1875–1925.* Marburg: Jonas.

Meyers, Amy R. W., and Margareth B. Pritchard, eds. 1998. *Empire's Nature: Mark Catesby's New World Vision.* Chapel Hill and London: University of North Carolina Press.

Miller, David, and Peter Reill, eds. 1996. *Visions of Empire: Voyages, Botany and Representations of Nature.* Cambridge: Cambridge University Press.

Minter, Sue. 1990. *The Greatest Glass House: The Rainforests Recreated.* London: HMSO.

Monmonier, Mark S. 1999. *Air Apparent: How Meteorologists Learned to Map, Predict, and Dramatize Weather.* Chicago: University of Chicago Press.

Monti, Nicholas. 1987. *Africa Then: Photographs, 1840–1910.* London: Thames & Hudson.

Moravia, Sergio. 1972. "Philosophie et médecine en France à la fin du XVIIIe siècle." *Studies in Voltaire and the Eighteenth Century* 39:1089–151.

Morgan, Susan. 1996. *Place Matters: Gendered Geography in Victorian Women's Travel Books about Southeast Asia.* New Brunswick, NJ: Rutgers University Press.

Moureau, F., ed. 1995. *L'oeil aux aguets ou l'artiste en voyage.* Paris: Klincksiek.

Naylor, Simon. 2000. "'That Very Garden of South America': European Surveyors in Paraguay." *Singapore Journal of Tropical Geography* 21:48–62.

Nicholson, M. 1990. "Alexander von Humboldt and the Geography of Vegetation." In *Romanticism and the Sciences,* edited by A. Cunningham and N. Jardine. Cambridge: Cambridge University Press.

Noblett, Bill. 1985. "Dru Drury's 'Directions for Collecting Insects in Foreign Countries.'" *Amateur Entomologist's Society Bulletin* 44:170–78.

Nordstrom, Alison Devine. 1990. "Images of Paradise: Photographs of Samoa, 1880–1930." Master's thesis, University of Oklahoma.

———. 1991. "Early Photography in Samoa: Marketing Stereotypes of Paradise." *History of Photography* 15:272–86.

Osborne, Michael. 2001. "Acclimatizing the World." *Osiris* 15:135–51.

Otis, Laura. 2000. *Membranes: Metaphors of Invasion in Nineteenth-Century Literature, Science, and Politics.* Baltimore and London: Johns Hopkins University Press.

Outram, Dorinda. 1996. "New Spaces in Natural History." In *Cultures of Natural History,* edited by Nicholas Jardine, James Secord, and Emma Spary. Cambridge: Cambridge University Press.

———. 1999. "On Being Perseus: New Knowledge, Dislocation and Enlightenment Exploration." In *Geography and Enlightenment,* edited by David N. Livingstone and Charles W. J. Withers. Chicago: University of Chicago Press.

Pagden, Anthony. 1993. *European Encounters with the New World.* New Haven, CT, and London: Yale University Press.

Paradis, James. 1985. "Darwin and Landscape." In *Victorian Science and Victorian Values: Literary Perspectives,* edited by James Paradis and Thomas Postlewait. New Brunswick, NJ: Rutgers University Press.

Paris, E. 1996. "L'Époque brésilienne de Fernand Braudel (1935–1937)." *Storia della Storiografia* 30:56–68.

Peard, Julyan. 2000. *Race, Place, and Medicine: The Idea of the Tropics in Nineteenth-Century Brazilian Medicine.* Durham, NC: Duke University Press.

Pick, Daniel. 1989. *Faces of Degeneration: A European Disorder, c. 1848–c. 1918.* Cambridge: Cambridge University Press.

Pickering, Jane. 1998. "William John Burchell's Travels in Brazil, 1825–1830, with Details of the Surviving Mammals and Bird Collections." *Archives of Natural History* 25, no. 2:237–66.

Pinney, Christopher, and Nicolas Peterson, eds. 2003. *Photography's Other Histories.* Durham, NC, and London: Duke University Press.

Poole, Deborah. 1997. *Vision, Race and Modernity: A Visual Economy of the Andean Image World.* Princeton, NJ: Princeton University Press.

Porter, Roy. 1995. "Medicine and the Science of Man in the Eighteenth Century." In

Inventing Human Science, edited by Christopher Fox and Robert Wokler. Berkeley: University of California Press.

Pratt, Mary-Louise. 1992. *Imperial Eyes: Travel Writing and Transculturation.* London: Routledge.

Preston, Rebecca. 1999. "'The Scenery of the Torrid Zone': Imagined Travels and the Culture of Exotics in Nineteenth-Century British Gardens." In *Imperial Cities: Landscape, Display and Identity,* edited by Felix Driver and David Gilbert. Manchester: Manchester University Press.

Quilley, Geoff, and John Bonehill, eds. 2004. *William Hodges, 1744–1797: The Art of Exploration.* New Haven, CT: Yale University Press.

Quilley, Geoff, and Kay Dian Kriz, eds. 2003. *An Economy of Colour: Visual Culture and the Atlantic World, 1660–1830.* Manchester: Manchester University Press.

Raffles, Hugh. 2002. *In Amazonia: A Natural History.* Princeton, NJ: Princeton University Press.

Raj, Kapil. 2003. "Circulation and the Emergence of Modern Mapping: Great Britain and Early Colonial India, 1764–1820." In *Society and Circulation: Mobile People and Itinerant Cultures in South Asia, 1750–1950,* edited by C. Markovits, J. Pouchepadass, and S. Subrahmanyam. Delhi: Permanent Black.

Ralston, Caroline. 1977. *Grass Huts and Warehouses: Pacific Beach Communities of the Nineteenth Century.* Canberra: Australian National University Press.

Redfield, Peter. 2000. *Space in the Tropics: From Convicts to Rockets in French Guiana.* Berkeley: University of California Press.

Reingold, Nathan. 1964. "Two Views of Maury . . . and a Third." *Isis* 55, no. 3 : 370–2.

Rigby, Nigel. 1998. "The Politics and Pragmatics of Seaborne Plant Transportation, 1769–1805." In *Science and Exploration in the Pacific: European Voyages in the Southern Oceans in the Eighteenth Century,* edited by Margarette Lincoln. Woodbridge: Boydell Press and the National Maritime Museum.

Rix, Martin. 1981. *The Art of the Botanist.* Guildford and London: Lutterworth Press.

Rozwadowski, Helen. 1996. "Fathoming the Ocean: Discovery and Exploration of the Deep Sea, 1840–1880." PhD diss., University of Pennsylvania.

Rudwick, Martin. 1976. "The Emergence of a Visual Language for Geological Science, 1760–1840." *History of Science* 14 : 149–95.

———. 2000. "Georges Cuvier's Paper Museum of Fossil Bones." *Archives of Natural History* 27 : 51–68.

Ryan, James R. 1997. *Picturing Empire: Photography and the Visualisation of the British Empire.* London: Reaktion Books.

Salesa, Damon. 1997. "'Troublesome Half-Castes': Tales of a Samoan Borderland." PhD diss., University of Auckland.

Savage, Victor. 1984. *Western Impressions of Nature and Landscape in Southeast Asia.* Singapore: Singapore University Press.

Schaffer, Simon. 1992. "Self-Evidence." *Critical Inquiry* 18 : 327–62.

Secord, Anne. 2002. "Botany on a Plate: Pleasure and the Power of Pictures in Promoting Early Nineteenth-Century Scientific Knowledge." *Isis* 93, no. 1 : 28–57.

Sinavaiana, Caroline. 1992. "Where the Spirits Laugh Last: Comic Theatre in Samoa." In *Clowning as Critical Practice: Performance Humour in the South Pacific,* edited by William E. Mitchell. Pittsburgh: University of Pittsburgh Press.

Sioh, M. 1998. "Authorizing the Malaysian Rainforest: Configuring Space, Contesting Claims and Conquering Imaginaries." *Ecumene* 5:144–66.

Skidmore, T. E. 2003. "Lévi-Strauss, Braudel and Brazil: A Case of Mutual Influence." *Bulletin of Latin American Research* 22:340–49.

Slater, Candace. 2002. *Entangled Edens: Visions of the Amazon.* Berkeley: University of California Press.

Smith, Bernard. [1960] 1985. *European Vision and the South Pacific.* 2d ed. New Haven, CT: Yale University.

Solomon-Godeau, Abigail. 1989."Going Native." *Art in America* 77:118–29.

Spate, O. H. K. 1977. " 'South Sea' to 'Pacific Ocean': A Note on Nomenclature." *Journal of Pacific History* 12, no. 4:205–11.

Stafford, Barbara Maria. 10984. *Voyage into Substance: Art, Science, Nature and the Illustrated Travel Account, 1760–1840.* Cambridge, MA and London: MIT University Press.

———. 1993. "Presuming Images and Consuming Words: The Visualization of Knowledge from the Enlightenment to Post-modernism." In *Consumption and the World of Goods,* edited by John Brewer and Roy Porter. London: Routledge.

———. 1993. *Body Criticism: Imagining the Unseen in Enlightenment Art and Medicine.* Cambridge, MA: MIT Press.

Stafford, Robert. 1989. *Scientist of Empire: Sir Roderick Murchison, Scientific Exploration and Victorian Imperialism.* Cambridge: Cambridge University Press.

Staszak, Jean-François. 2003. *Géographies de Gauguin.* Paris : Editions Bréal.

Stearns, Raymond Phineas. 1970. *Science in the British Colonies of America.* Chicago and London: University of Illinois Press.

Stepan, Nancy Leys. 2000. "Tropical Modernism: Designing the Tropical Landscape." *Singapore Journal of Tropical Geography* 21, no. 1:76–91.

———. 2001. *Picturing Tropical Nature.* London: Reaktion Books.

Swainson, Geoffrey M., ed. 1992. *William Swainson, Naturalist and Artist.* Palmerston North: Swift.

Tagg, John. 1988. *The Burden of Representation: Essays on Photographies and Histories.* Basingstoke: Macmillan Education.

Taylor, David. 2000. "A Biogeographer's Construction of Tropical Lands: A. R. Wallace, Biogeographical Method and the Malay Archipelago." *Singapore Journal of Tropical Geography* 21:63–75.

Thomas, Nicholas. 1989. "The Force of Ethnology: Origins and Significance of the Melanesia/Polynesia Division." *Current Anthropology* 30:27–41.

———. 1991. *Entangled Objects: Exchange, Material Culture and Colonialism in the Pacific.* Cambridge, MA: Harvard University Press.

———. 1994. *Colonialism's Culture: Anthropology, Travel and Government.* Oxford: Polity.

———. 1996. " 'On the Varieties of the Human Species': Forster's Comparative Ethnology." In *Observations Made during a Voyage Round the World,* by J. R. Forster. Edited by Nicholas Thomas, Harriet Guest, and Michael Dettelbach. Honolulu: University of Hawai'i Press.

———. 1997. "Melanesians and Polynesians: Ethnic Typifications Inside and Outside Anthropology." In *In Oceania: Visions, Artifacts, Histories.* Durham, NC: Duke

University Press.

Thomas, Nicholas, and Diane Losche. 1999. *Double Vision: Art Histories and Colonial Histories in the Pacific.* Cambridge: Cambridge University Press.

Thomaz, Omar Ribeiro. 1996. "Do Saber Colonial ao Luso-Tropicalismo: 'Raça' e 'Nação' nas Primeiras Décadas do Salazarismo." In *Raça, Ciência e Sociedade,* edited by Marcos Chor Maio and Ricardo Ventura Santos. Rio de Janeiro: Fiocruz/ CBBB.

Tillotson, Giles. 2000. *The Artificial Empire: The Indian Landscapes of William Hodges.* Richmond, Surrey: Curzon Press.

Tobin, Beth Fowkes. 1999. *Picturing Imperial Power: Colonial Subjects in Eighteenth-Century British Painting.* Durham, NC, and London: Duke University Press.

———. 2002. "Tropical Bounty, Local Knowledge, and the Imperial Georgic." In *Monstrous Dreams of Reason: Body, Self, and Other in the Enlightenment,* edited by Mita Choudhury and Laura J. Rosenthal. Lewisburg, PA: Bucknell University Press.

Turnbull, Paul. 1999. "Enlightenment Anthropology and the Ancestral Remains of Australian Aboriginal People." In *Voyages and Beaches: Pacific Encounters, 1769–1840,* edited by Alex Calder, Jonathan Lamb, and Bridget Orr. Honolulu: University of Hawai'i Press.

Walcott, Derek. 1986. "Midsummer" (1984). In *Collected Poems, 1948–1984.* New York: Farrar, Straus & Giroux.

Walvin, James. 1997. *Fruits of Empire: Exotic Produce and British Taste, 1660–1800.* Basingstoke: Macmillan.

Wareham, Evelyn. 1999. "Developing Samoan Difference, 1900–1914." *Turnbull Library Record: Pacific Studies Issue* 32:57–74.

Williams, Francis Leigh. 1963. *Matthew Fontaine Maury, Scientist of the Sea.* New Brunswick, NJ: Rutgers University Press.

Withers, Charles W. J. 1995. "Geography, Natural History and the Eighteenth-Century Enlightenment: Putting the World in Place." *History Workshop Journal* 39:137–63.

Worboys, Michael. 1976. "The Emergence of Tropical Medicine: A Study in the Establishment of a Scientific Specialty." In *Perspectives on the Emergence of Scientific Disciplines,* edited by G. Lemaine, R. MacLeod, M. Mulkay, and P. Weingart. Chicago: A. Hine.

———. 1993. "Tropical Diseases." In *Companion Encyclopedia of the History of Medicine,* edited by W. F. Bynum and Roy Porter. 2 vols. London and New York: Routledge.

Notes on Contributors

DAVID ARNOLD is professor of the history of South Asia at the School of Oriental and African Studies, University of London. His books include *Colonizing the Body* (1993), *The Problem of Nature* (1996), and *Gandhi* (2001). *The Tropics and the Traveling Gaze: India, Landscape and Science, 1800–1856* is due to be published in 2005.

LEONARD BELL is associate professor of art history at the University of Auckland. He is the author of *Colonial Constructs: European Images of Maori, 1840–1914* (1992). He has published numerous articles on visual imagery, empire, and cross-cultural interactions in the Pacific.

D. GRAHAM BURNETT is assistant professor of history at Princeton University. His book *Masters of All They Surveyed* was published by the University of Chicago Press in 2000. He is a coeditor of volume 4 of *The History of Cartography* (forthcoming).

DENIS COSGROVE is Alexander von Humboldt Professor of Geography at the University of California, Los Angeles. His recent books include *Apollo's Eye: A Cartographic Genealogy of the Earth in the Western Imagination* (2001) and, as editor, *Mappings* (1999).

MICHAEL DETTELBACH is a fellow of the Center for the History and Philosophy of Science at Boston University. He has published on the work of Alexander von Humboldt, including an introduction to a reprint of *Cosmos* (1997), and has coedited J. R. Forster's *Observations Made during a Voyage Round the World* (1996).

STARR DOUGLAS recently completed a PhD dissertation entitled "Natural History, Improvement and Colonisation: Henry Smeathman and Sierra Leone in the Late Eighteenth Century" at Royal Holloway, University of London (2004).

FELIX DRIVER is professor of human geography at Royal Holloway, University of London. He is the author of *Geography Militant: Cultures of Exploration and Empire* (2001) and coeditor of *Imperial Cities: Landscape, Display and Identity* (1999).

ROD EDMOND is professor of modern literature and cultural history at the University of Kent. He is the author of *Affairs of the Hearth: Victorian Poetry and Domestic Nar-*

rative (1988) and *Representing the South Pacific: Colonial Discourse from Cook to Gaugin* (1997) and coeditor of *Islands in History and Representation* (2003).

CLAUDIO GREPPI is professor of geography at the University of Siena. His works include *Intorno a Humboldt* (1996), *Paesaggi dell'Appennino toscano* (1990–93), and, as editor, *L'invenzione del Nuovo Mondo: Critica della conoscenza geografica* (1992), a translation of the first two volumes of Humboldt's *Examen critique de l'histoire de la géographie due Nouveau Continent*.

PETER HULME is professor of literature at the University of Essex. He is the author of *Colonial Encounters* (1986) and *Remnants of Conquest: The Island Caribs and Their Visitors, 1877–1998* (2000) and coeditor of *The Tempest and Its Travels* (2000) and *The Cambridge Companion to Travel Writing* (2002).

LUCIANA MARTINS is lecturer in Luso-Brazilian studies at Birkbeck, University of London. She is the author of *O Rio de Janeiro dos Viajantes: O Olhar Britânico, 1800–1850* (2001) and articles on the visual culture of tropical travel.

Index